SANDALWOOD CULTIVATION AND UTILISATION

Dr. Vu Van Thoai

Dr. Ashutosh Srivastava

M. Srinivasa Rao

walnutpublication
.com

INDIA • UK • USA

Copyright © Authors, 2020

Paperback ISBN: 978-1-954399-13-6

Hardback ISBN: 978-1-954399-14-3

eBook ISBN: 978-1-954399-15-0

First Published in December 2020

Published by Walnut Publication
(an imprint of Vyusta Ventures LLP)

www.walnutpublication.com

USA

6834 Cantrell Road #2096, Little Rock, AR 72207

India

#722, Esplanade One, Rasulgarh, Bhubaneswar – 751010

#55 S/F, Panchkuian Marg, Connaught Place, New Delhi - 110001

UK

International House, 12 Constance Street, London E16 2DQ

Printed in India by KSRK Impressions

Khasra No.1186/5, Jain Complex, Sirsakhurd, Durg – 491001

FOREWORD

Santalum album L. (East Indian Sandalwood) is a medium sized semi-parasitic tree which can reach to a height of 33 to 66 feet in height. In India, the tree is called as miracle tree because of its immense uses in trade and traditional form of medicine. As it is semi parasitic in nature it prefers to "steal" nutrients from the roots of nearby plants by using tube-like structures (modified roots) called haustoria. By doing so, sandalwood extracts the resources of the host plants. Essential oils of sandalwood have antispasmodic (relieve spasm of muscles), antibacterial (kill bacteria) and antiviral (kill viruses) properties. Sandalwood is used in treatment of inflammation, skin disorders, rashes, cough, fever, flu, hypertension and restlessness. Essential oils are used to improve functioning of liver, heart and stomach and to strengthen the gums of teeth and muscles. Hardwood of sandalwood was used for building the temples and statues of Hindu gods and elephants in India and other South East Asian countries. Due to belief that sandalwood plays role in reincarnation, essential oils are often used in the funeral rituals and for embalming of the dead bodies.

Under natural conditions Sandalwood is ready for the harvest at the age of 30 to 60 years when the heartwood is matured and sufficient oil has been developed. But with the development of modern cultivation technologies it attains a harvestable stage in 15 years. This makes its commercial cultivation possible on the farmer's field. Another major reason and need for its commercial cultivation is the fact that natural populations of Sandalwood are not sufficient to meet the current world demand. As per an estimate the current requirement of Sandalwood is around 40,000 MT per year whereas supply is not even 100 MT. This scenario presents a great opportunity for large scale cultivation of sandalwood and its commercial utilisation. Vietnam Sandalwood Institute has been dedicated to this cause and has started cultivation of Sandalwood in over 5000 acres in Vietnam. This book is a collection of information collected from India, Vietnam, Australia and elsewhere on its cultivation practices, scope and utilisation. This can serve as a ready reference for people who aim to grow it on commercial scale in Vietnam and other ASEAN countries.

Dr Vu Van Thoai
Chairman,
Vietnam Sandalwood Institute

Table of Contents

Chapter 1: General introduction   1-24

1.1. Economic value of sandalwood

1.2. Botany, History and distribution of sandalwood

    1.2.1 Botany

    1.2.2 History of sandalwood

    1.2.3 Distribution of Sandalwood in India and World

        1.2.3.1 Distribution of Sandalwood in India

        1.2.3.2 Distribution of Sandalwood in World

        1.2.4 Global production and International Market

Chapter 2: Classification, Genetics, Tree   25-33
improvement and breeding of Sandalwood

2.1. Classification of Sandalwood

    2.1.1 Botanical Classification of sandalwood

    2.1.2 Commercial classification of sandalwood sold in India

2.2 Genetics, tree improvement and breeding of Sandalwood

2.2.1 Genetics of Sandalwood

2.2.2 Genetics and Karyotype analysis of Sandalwood

2.2.3 Tree improvement and Breeding of Sandalwood

Chapter 3: Silviculture and Propagation of Sandalwood   **34-50**

3.1. Climate and Soil

3.2. Propagation methods

    3.2.1 Seed propagation

        3.2.1.1 Earlier work done on seed propagation

        3.2.1.2 Collection and processing of seeds

        3.2.1.3 Seed bed preparation, sowing and maintenance

        3.2.1.4 Transplanting of seedlings

        3.2.1.5 Nursery hosts of Sandalwood

        3.2.1.6 Maintenance of seedlings

    3.2.2 Vegetative propagation

        3.2.2.1 Marcotting

        3.2.2.2 Grafting

        3.2.2.3 Tissue culture

Chapter 4: Sandalwood nursery establishment and management   51-54

4.1 Factors to be considered before establishing nursery

4.2 Design and layout of the nursery

4.3 Nursery management

Chapter 5: Establishment of Sandalwoodood plantations   55-69

5.1. Selection of site

    5.1.1 Biophysical considerations for site selection

    5.1.2 Socio-economic factors to be considered for site selection

5.2. Land preparation

    5.2.1 Methods of site preparation

        5.2.1.1 Soil and water conservation structures

            5.2.1.1.1 Earthern bunds

            5.2.1.1.2 Graded bunds

            5.2.1.1.3 Vegetative hedges

            5.2.1.1.5 Trenches

            5.2.1.1.6 Strip terrace

            5.2.1.1.7 Moisture conservation pits

5.2.1.1.8 Farm ponds for water conservation

5.2.2 Plantation designing and planting

5.2.2.1 Marking and pitting

5.2.2.2 Spacing in the sandalwood plantation

5.2.2.3 Grading and transportation of seedlings

5.3 Planting of seedlings and the hosts

Chapter 6: Sandalwoodood plantation management   **70-88**

6.1. Growth management plan

6.2 Silvicultural operations

6.2.1 Tree shaping

6.2.1.1 Formative pruning

6.2.1.2 Form pruning

6.2.2 Weeding

Chapter 7: Pests and diseases of sandalwood in nurseries and plantations and their management   **89-119**

7.1 Insect pests of nursery

7.2 Diseases of nursery

7.3. Insect pests of plantation

7.4. Diseases of plantations

7.5 Nutrient Deficiencies of plantations

Chapter 8: Heartwood formation and harvesting   **120-123**

8.1. Heartwood formation in vascular plants

8.2. Sandalwood harvesting

    8.2.1 Preliminary processing

    8.2.2 Sandalwood processing for carving and agarbatti

References   **124-138**

CHAPTER 1

GENERAL INTRODUCTION

1.1. Botany, History, distribution of sandalwood

1.1.1 Botanical description

Santalum album is a small evergreen tree that grows to 4 m in Australia, but in India it is much larger and can grow to a height of 20 m girth of up to 2.4 m, with slender drooping branchlets. Bark is tight, dark brown, reddish, dark grey or nearly black, smooth in young trees, rough with deep vertical cracks in older trees, red inside. Leaves thin, usually opposite, ovate or ovate elliptical, 3-8 x 3-5 cm, glabrous and shining green above, glaucous and slightly paler beneath; tip rounded or pointed; stalk grooved, 5-15 cm long; venation noticeably reticulate. Flowers purplish-brown, small, straw coloured, reddish, green or violet, about 4-6 mm long, up to 6 in small terminal or axillary clusters, unscented in axillary or terminal, paniculate cymes. Fruit a globose, fleshy drupe; red, purple to black when ripe, about 1 cm in diameter, with hard ribbed endocarp and crowned with a scar, almost stalk less, smooth, single seeded. The generic name is derived from the Greek 'santalon' meaning 'sandalwood', and the species name from the Latin 'albus' meaning 'white'(Orwa *et al* 2009).

Sandalwood is the fragrant heartwood of species of genus *Santalum* (family –Santalaceae). In India, the genus is represented by *Santalum album Linn*. Its wood, known commercially as "East Indian Sandalwood" and essential oil from it as "East Indian Sandalwood Oil" are among the oldest known perfumery materials. There are around 18 sandalwood species found all over the world belonging to the genus *Santalum* which are; *S. freycinetianum, S. haleakalae, S. ellipticum, S. peniculatum, S. pyrularia, S. involutum, S. boninese, S. insulare, S. austrocaledonicum, S. yasi, S. macgregorii, S. accuminatum, S. murrayanum, S. obtusifolium, S. lanceolatum, S. fernandezianum, S. salicifolium and S. spicatum.* All the sandalwood species are identified to be obligate wood hemi-parasites which mean they absorb certain nutrients such as phosphates and nitrates from the host trees via root connections called haustoria (Subasinghe 2013).

Flower panicles of *Santalum album* appear from March to April in India, and fruits ripen in the October November; in Australia flowers appear in December to January and also June to August, and mature fruit is available from June to September. The species is spread rapidly through seed dispersal by birds, which feed on the outer fleshy pericarp. Viable seed production occurs when the tree is over 5 years old (Orwa *et al* 2009).

1.1.2 History of Sandalwood

Historical review reveals that sandalwood has been referred to in Indian mythology, folklore and ancient scriptures. It is generally accepted that sandal is indigenous to peninsular India as its history of recorded occurrence dates back to at least 2500 years However, there are varied views regarding the origin of Sandalwood some of them who consider East Timor as the birth place of sandalwood and they opine that it was introduced in India about 2000 years back (Roxburgh, 1820; Sprague and Summerhayes, 1927; Fischer, 1928, 1938; Tuyama, 1939; St. John, 1947; Thirawat, 1955; Shetty, 1977; Mc Kinnell, 1990; Rai, 1990). India has been the traditional leader of sandalwood oil production for perfumery and pharmaceuticals. The aroma of the oil and the wood is esteemed by people belonging to three major religions of the world – Hinduism, Buddhism and Islam. The ancient Egyptians imported the wood and used it in medicine, for embalming the dead and in ritual burning to venerate the gods. Long before the reorganization of states in India, It is customary in certain communities among the Hindus to put a piece of sandalwood in the funeral pyre. The beige-coloured paste of sandalwood is applied on the forehead and other body parts, especially by devotees of God Krishna (Vaishnavites) and for ritual bathing of Hindu gods. (Arun Kumar *et al* 2012)

Sandalwood Tree

Leaves

Flowers

Unripened fruits

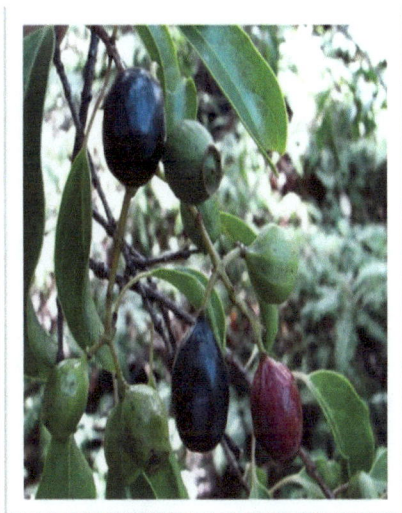

Ripened fruits

1.1.3 Distribution of Sandalwood

1.1.3.1 Distribution in India

In the past, it naturally occurred in peninsular India, but subsequently it has been introduced in other parts too. It generally occurs in the dry deciduous forests of Deccan Plateau at the edge of the Western Ghats Range. Apart from Karnataka, sandal trees are also distributed in states of Andhra Pradesh, Kerala, Maharashtra, Madhya Pradesh, Odisha, Rajasthan, Uttar Pradesh, Bihar and Manipur. In Kerala, Sandal is spread over 15 km^2 mainly in Marayur in Idukki district, Wayanad district and Thenmalai in Kollam district (Srimathi *et al.*, 1995). Over 90% of sandal is distributed in Karnataka (5245 km^2) and Tamil nadu (3045 km^2), and rest in Andhra Pradesh (175 km^2), Orissa (35 km^2), Madhya Pradesh (33 km^2) and Maharashtra (84 km^2) (Jeeva *et al.*, 1998). In the current scenario, after the relaxation of policies by Karnataka and Tamil Nadu Governments sandalwood is being cultivated elsewhere outside its traditional natural zone. Sporadic occurrence of Sandalwood is reported from states of Himachal Pradesh, Punjab,

Assam, Madhya Pradesh which is attributed to the practice of gifting seedlings by erstwhile Royal Family of Mysore to other Royal families of pre-independent India.

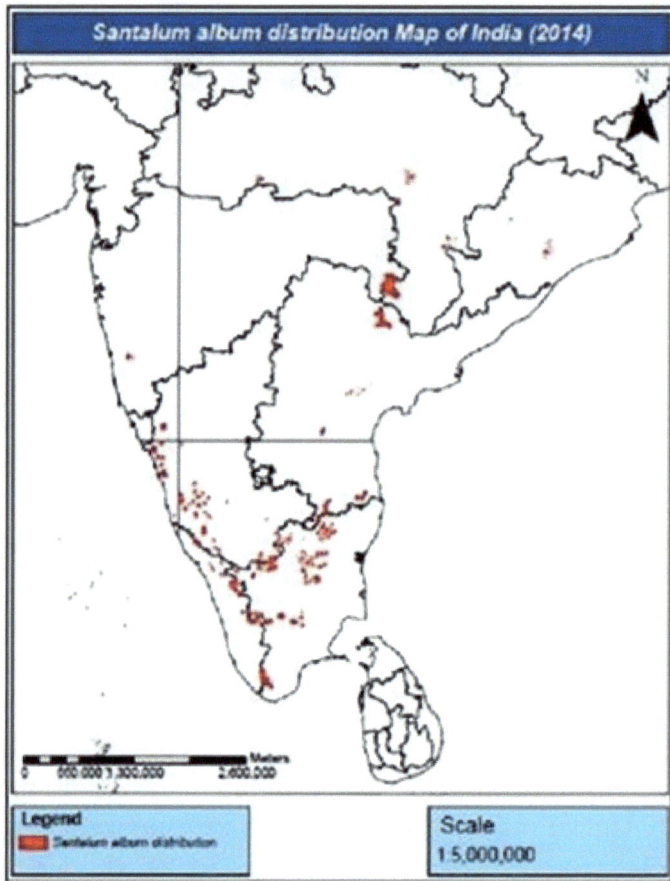

Source: Rajan and Jayalakshmi (2017)

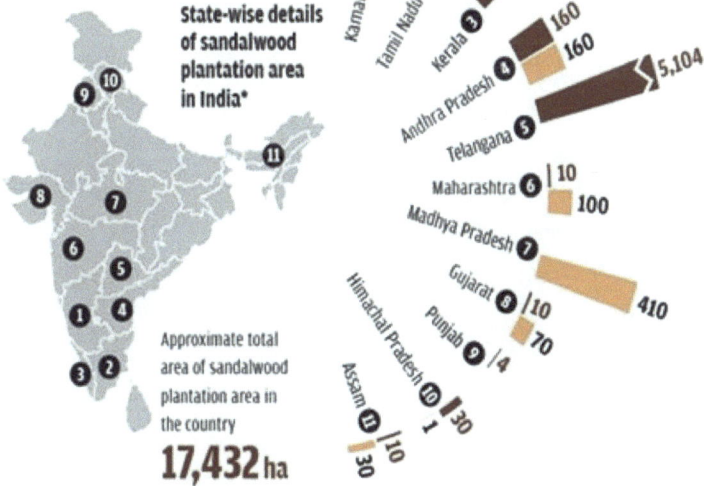

Rise of the royal tree

Total land under sandalwood cultivation in India is increasing by 600 hectares annually

State-wise details of sandalwood plantation area in India*

Approximate total area of sandalwood plantation area in the country

17,432 ha

Legend:
- Government area (in ha)
- Private area (in ha)

Karnataka ❶ 10,400 / 375
Tamil Nadu ❷ 410
Kerala ❸ 80 / 68
Andhra Pradesh ❹ 160 / 160
Telangana ❺ 5,104
Maharashtra ❻ 10 / 100
Madhya Pradesh ❼
Gujarat ❽ 10 / 410
Punjab ❾ 4 / 70
Himachal Pradesh ❿ 1 / 30
Assam ⓫ 10 / 30

Source: Pallavi, A. (2018)

1.1.3.2 Distribution of Sandalwood in World

The genus '*Santalum*' grows naturally throughout the Pacific and Eastern Indian Ocean regions. Sandalwood trees are evergreen ranging in size from tall shrubs to large trees. They grow in a variety of climates ranging from Australian desert to sub-tropical New Caledonia and at elevations from sea level to 5000 meters. It is distributed in parts of Malaysia, Australia, New Zealand and Polynesia extending to the Hawaiian Archipelago and Juan Fernandez Islands distributed between 30°N and 40° S (Ananthapadmanabha and Gowda 2011). It is also localized in some part of the mid-western development region in Nepal the species has been reported from

Gorkha (Majupuria & Joshi, 1988) and Pyuthan Provinces of Nepal (Krishna Bahadur, 2019). It is largely distributed in Sri Lanka and other parts of the south-eastern Asia (Brandis, 1978; Hari Shankar Lal *et al.* 2014). However, it is supposed to be introduced there from Java, Indonesia (Neil, 1990; Gusti Ngurah Sudiana, 2013). In Indonesia *Santalum album* has been reported to be growing endemically in the Province of East Nusa Tenggara (NTT) in the islands of Timor, Sumba, Alor, Solor, Pantar, Flores, Rote, and other islands. Besides in NTT, sandalwood is also found at Gunung Kidul, Imogiri, Kulon Progo at the Special Province of Yogyakarta (DIY), Bondowoso (East Java), and the island of Sulawesi (Yani *et al* 2010). Recently, commercial plantations of sandalwood are being raised in Inida, Australia, Indonesia and China. In Vietnam , models of sandalwood grown with variety of permanent hosts crops like orange, grape fruit, Jackfruit, avocado, durian, coffee etc has been growing with success.

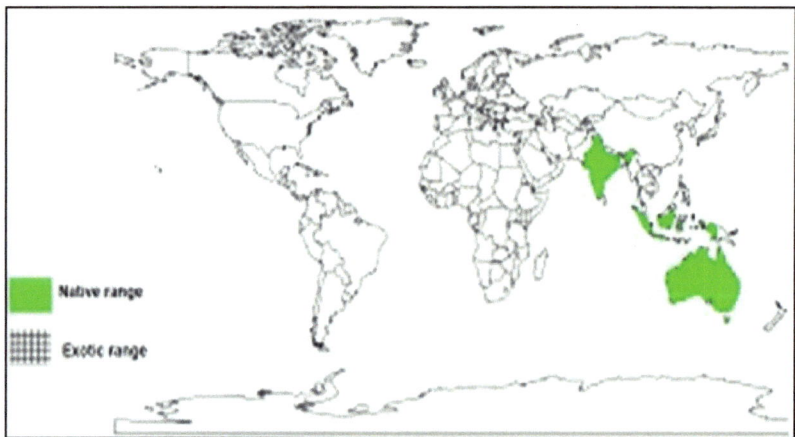

Source: Orwa et al (2009)

1. AUSTRALIAN SANDALWOOD

Sandalwood is an integral part of the history and ecology of Australia. There are six species of *Santalum* grown naturally, *Santalum spicatum*, *S. album*, *S.lanceolatum*, *S.murrayanum*, *S. accuminatum* and *S. obtusifolium*. Out of six species only three species, *S. spicatum*, *S.album*, and *S.lanceolatum* yield commercially important scented heartwood and oil. *S. spicatum*, popularly known as West Australian sandalwood is also one of the important species producing scented heartwood and oil. Wood has been exported to different countries since 1844. The oil has a different chemical composition and has better medicinal properties. The oil is used in perfumery, incense sticks, soaps and toiletries and aromatherapy. The wood is also used in carving. Recently, it has become an alternate source of essential oil for sandalwood based industries. Wood production is on the decline, and adequate conservative methods to limit the annual harvest are being taken by the government. Since 2000, large plantation activities are being taken up by corporate bodies to meet the growing demand. *Santalum lanceolatum* is the second major type of sandalwood native to Australia; the heartwood contains fewer odours and yields comparatively less oil; the oil mixes well with other sandalwood oil. The scented oil is used in perfumery industry. *S.accuminatum* is also a native species of Australia and the trees are cultivated for fruits rather than oil. The timber is hard, usually used for furniture making and is not scented. *S.murrayanum* is a small shrub or a small tree growing to 4 meters with spreading branches. The wood is not perfumed and bark is more or less smooth. The tree has no commercial value. *S.obtusifolium* is a small shrub growing up to 2.6 m., the wood is not perfumed and has no commercial value (Ananthapadmanabha and Gowda, 2011).

2. PACIFIC ISLAND SANDALWOOD

S. austrocaledonicum, is an important species growing in New Caledonia and Vanuatu islands, scented heartwood is used for making beautiful handicraft items and the oil produced from heartwood is highly scented and are sold as New Caledonian oil or New Caledonian sandalwood oil. The oil is highly priced and very much sought after in the perfume industries. The species has been over exploited commercially, and annual harvest exceeds 100 tonnes.. Sandalwood contributes substantial forest revenue to the island. Attempts are being made to conserve the species in different pacific islands. *S. yasi* grows in Fiji and in Tonga islands, its scented heartwood yield high quality sandalwood oil. Fijians and Tonganians scent coconut oil during marriage ceremonies, in which the bride and bridegroom are dusted with powdered heartwood. The fruit and flowers are similar to that of *S. album*, but the leaves are oblong. The heartwood is yellow and highly scented. It has been reported that native Fijian species *S.yasi* is able to naturally hybridise with *S. album*. The seeds collected from *S.yasi* trees in mixed plantation with *S. album* gave 20% hybrid seeds. Hybrid sandalwood seedlings showed growth rate almost twice to that of *S.yasi*, achieving over 2.5 m height in two years. A more detailed studies need to be carried out on artificially producing this hybrid using superior parents of both the species. The oil is similar to that *S.album* and the oil yield is about 7 per cent in the natural populations (Ananthapadmanabha and Gowda, 2011).

3. HAWAIIAN SANDALWOOD

Four species of sandalwood are found distributed on the Island. They are small shrubs or trees typically 5 to 10 m or larger at maturity. All species exhibit considerable morphological variations and numerous traditional varieties are recognized. Wood is traditionally used in the Pacific islands for carvings, cultural uses and is burnt as insect repellent. All species are attractive, especially when in flower and suitable for home gardens. 1. *Santalum ellipticum* 2. *Santlum freycinetianum* 3. *Santalum haleakalae* 4. *Santalum paniculatum* Sandalwood has strong history and is particularly culturally attached to different religions. The product derived from it has been in use for several centuries. Other species of Santalum are only of academic interest since their production and contribution to the essential oil industry is limited;

however, they are the important species of the region contributing to the cultural heritage and use, but due to over exploitation of the species, many have disappeared or in the endangered list. *Santalum macgregori*, called Papua New Guinea sandalwood, is another species producing scented heartwood and oil used in perfumery.

Non-Santalum species that are referred to as "sandalwood" include:

Amyris balsamifera – Amyris, or "West Indian Sandalwood" – a term little used today.

Osyris lanceolata and *Osyris speciosa* – both referred to as "African sandalwood".

Brachyleana hutchinsii – Muhuhu, sometimes also called "African sandalwood".

Erythroxylum monogynum – "Bastard sandal"

1.2. Economic value of sandalwood

Sandalwood has immense value in trade due to its versatile uses. Over the years, due to dwindling supply, the prices are increasing every year (Viswanath *et al*, 2014). In Government auctions the current value of heartwood is INR 7000-8000/kg in India. However, the private parties buy it at much lower prices of INR 2800-3000 per kg. The main reason behind this is long procedure of over 1 year involved in selling the product to government agencies and extreme classifications the government agencies carry out in evaluation of the sandalwood (Personal Communication).

International prices are 20% higher but sandalwood export in India is banned in log form. High quality sandalwood oil is sold at the rate of Rs.2,50,000/ kg by the private parties while the government agencies like Cauvery (A Government of Karnataka Undertaking) sell it at a very high value of Rs 6.0 lakhs per kg in retail market (Personal Communication). In nature, very old trees (>50 year old trees) yield an average of 60 kg of oil is obtained per ton of heartwood. However, scientifically grown plantation trees of 15-20 year old yield about 25-

30 kg of oil per ton. The price at the international market is about 15 to 20 % higher than in the domestic market.(Gowda and Ananthapadmanabha, 2011).

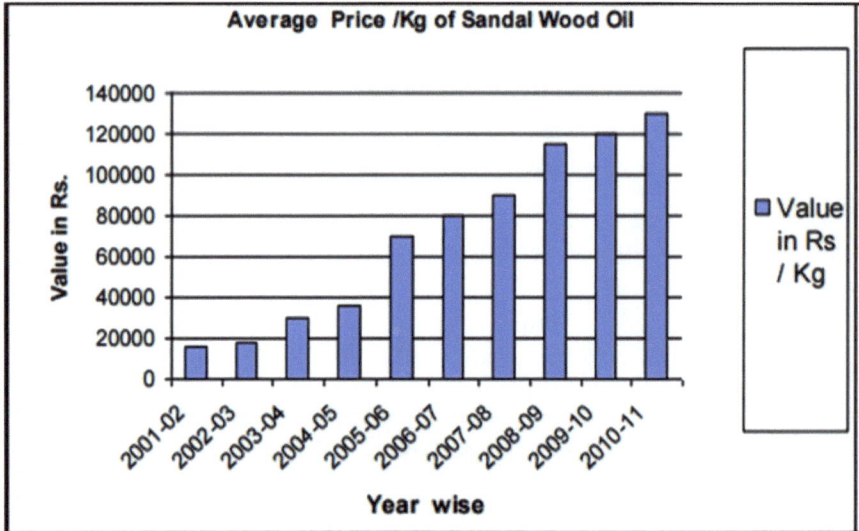

Source: (Gowda 2011)

1.2.1.) Different types of *Santalum* species found in World (Wikipedia, 2020)

Name of Species	Image	Distribution
S.freycinetianum		Endemic to Hawaiian Islands
S.haleakalae		Endemic to Hawaiian Islands
S.ellipticum		Endemic to Hawaiian Islands

1.2.2.) Different types of *Santalum* species found in World (Wikipedia, 2020)

Name of Species	Image	Distribution
S.peniculatum		Endemic to Hawaiian Islands (Huluhulu)
S.pyrularia		Hawaiian Islands and Papua New Guinea
S.involutum		Northern, southern, and western regions of Kaua`I region of Hawaii Islands Courtsey (Keneth R Wood)

1.2.3.) Different types of *Santalum* species found in World (Wikipedia, 2020)

Name of Species	Image	Distribution
S.boninese		(Nakai) Tuyama (Bonin Islands, Japan)
S.insulare		Southeast Polynesia (Cooks - French Polynesia)
S.austrocaledonicum		Endemic to New Caledonia (Grande-Terre, Isle of Pines, Loyalty Islands) and Vanuatu, Courtsey: Jean-François Butaud

1.2.4.) Different types of *Santalum* species found in World (Wikipedia, 2020)

Name of Species	Image	Distribution
S.yasi		Fiji, Niue and Tonga Islands of South Pacific Ocean
S.macgregorii		Papua New Guinea
S.accuminatum		South West and South East Australia

1.2.5.) Different types of *Santalum* species found in World (Wikipedia, 2020)

Name of Species	Image	Distribution
S.murrayanum		Western Australia
S.obtusifolium		Victoria, New South Wales and Queensland, Australia
S.lanceolatum		Western New South Wales, Australia

17

1.2.6.) Different types of *Santalum* species found in World (Wikipedia, 2020)

Name of Species	Image	Distribution
S.fernandezianum	© NATHAN/1999. Paris, France.	Endemic to Juan Fernández Islands off the coast of Chile.
S.salicifolium	Not available	Hawaii
S.spicatum		Australia

Common name	Botanical name	Grows in	Current oil exports	Sustainability & crop management
Indian Sandalwood	Santalum album	India*	Minimal	Very poor
		Sri Lanka*	Prohibited	Very poor
		Indonesia*	Minimal	Very poor
		Philippines*	None	Very poor
		China*	Not known	Not known
		Australia	Substantial	Very good
Australian Sandalwood	Santalum spicatum	Australia	Substantial	Good
Northern Queensland Sandalwood	Santalum lanceolatum	Australia	Limited	Poor
New Caledonian Sandalwood	Santalum austrocaledonicum	New Caledonia	Limited	Poor
Vanuatu Sandalwood		Vanuatu	Limited	Moderate
Hawaiian Sandalwood	Santalum paniculatum	Hawaii	Limited	Very poor

* Classified as "vulnerable" by the IUCN http://www.iucnredlist.org/details/31852/0

Different *Santalum* species occurring in different parts of the world have some characteristic features of their own and are culturally attached to the region.

Constituent	S. album	S. spicatum	S. austrocal.	S. paniculatum
(Z)-a-Santalol	41-55*	15-25*	38-46	>34
(Z)-b-Santalol	16-24*	5-20*	13-19	11-13
(Z)-a-Bergamotol	4-6	2-10*	5-8	>3.8
(Z)-Lanceol	1.5-3.5	1-10*	5-15	2-3
(E,E)-a-Farnesol	0.2-0.3	2.5-15*	0.9-1.7	>1.3
(Z)-Nuciferol	1.1-3.4	4.4-6.4	1.5-2.1	3-5
(Z)-b-Curcumen-12-ol	0.8-2	5.1-7.8	1.1-1.8	1.6-2.2

* International Organization for Standardization (ISO) standard. All other data derive from Braun et al 2014 or De Groot & Schmidt 2016

1.3 Sandalwood uses

1.3.1. Leaves

Active bio-components	Effect on human body
Urosilic acid	It is reported to be able to reduce fat accumulation and increase muscle mass gain when in a fed state, and to induce fat burning and preserve muscle mass when in a fasted state.
Catechin	Catechin is reported to have many beneficial properties for human health such as anticancer, anti-obesity, anti-diabetic, anti-cardiovascular, anti-infectious, hepato-protective, and neuroprotective effects.
Vitexin	Vitexin (5, 7, 4-trihydroxyflavone-8-glucoside) has recently received increased attention due to its wide range of pharmacological effects, including but not limited to anti-oxidant, anti-cancer, anti-inflammatory, anti-hyperalgesic, and neuroprotective effects.
Iso vitexin	Reported to positively affect high-density lipoprotein (HDL) cholesterol, blood pressure, and platelet function.
Shikimic acid	Pharmaceutical applications of SA include its use as an antipyretic, antioxidant, and anticoagulant, antithrombotic, anti-inflammatory, and analgesic agent. It is also an efficient inhibitor of the surface protein neuraminidase (NA) enzyme of the seasonal influenza virus types A and B.
Gama amino butyric acid (GABA)	Nerve relaxant, induces sleep, and cures stress related disorders.

1.3.2. Seed Oil

Active bio-components	Effect on human body
Xymnemic acid	Vasco dilation, Nerve protector, anti-inflammatory, anti-oxidant, anti-ageing, anti-cancerous.
Oleic acid	It has beneficial effect on cancer, autoimmune and inflammatory diseases, besides its ability to facilitate wound healing. It imparts pharmacological properties to the seed oil and can be structurally modified to yield powerful pain relaxant. Anti-viral by controlling the multiplication of HSV-1 by modulating liver's Gluthatione-S-transferase and Sulphydryl.
Linoleic acid	Anti-hyperlipidemia action
Stearic acid	Anti-hyperlipidemia action

1.3.3 Seed Cover

Active bio-components	Effect on human body
Betulinic acid	It is reported to be an effective anti proliferative and apoptosis-inducing property. Betulinic acid has been demonstrated to kill several cancer types including melanoma. Betulinic acid was also found to be effective at preventing small and non–small cell lung, ovarian, cervical, and head and neck carcinomas

1.3.4. Bark

Active bio-components	Effect on human body
Triterpenoid (3-beta-yl-palmitate)	Insect population controlling property by interfering and the development of pupae of the insects and producing non-functional sterile adults.

1.3.5. Wood Oil

Active bio-components	Effect on human body
Alpha-santalol	It is renowned for its sublime, aphrodisiac, diuretic, and urinary antiseptic properties.
	Also Reported to have chemo preventive effects on skin cancer, the antitumor and cancer preventive properties of alpha-santalol have been shown to involve cell death induction through apoptosis and cell cycle arrest in various cancer models. Also reported to have anti-hyperglycemic, anti-psychotic, anti-bacterial and anti-inflammatory properties. Provides pleasant odor to the oil.
Beta-santalol	Fragrant compound of the oil which works in combination with alpha-santalol for above therapeutic activities on human body. Effects male sexual health and functions.
Bergamotol	Effective in breast cancer.

1.2.4 Global Production and International Market:

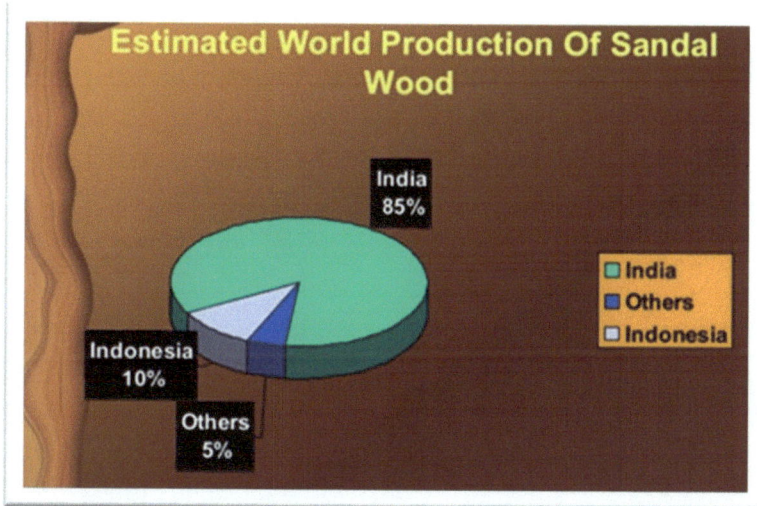

Estimated World Production Of Sandal Wood

India 85%

Indonesia 10%

Others 5%

India
Others
Indonesia

Source: Gowda (2011)

Indian sandalwood is prized for being the most potent, with about 90% content of the active ingredient santalol. In contrast, Australian sandalwood oil has only 38-39% santalol content. Not surprisingly, India dominates production worldwide. Nearly 85% of the supply of Indian sandalwood comes from the southern states of Karnataka, Tamil Nadu and Kerala.

Producing commercially valuable sandalwood with high levels of fragrance oils requires *Santalum* trees to be a minimum of 15 years old (*S. album*) the age at which they could be harvested .Yield of oil tends to vary depending on the age and location of the tree; usually, the older trees yield the highest oil content and quality. Australia is currently, the largest producer of *S. album* by 2018 (Anon 2018). The majority grown around Kununurra, Western Australia. Western Australian sandalwood is also grown in plantations in its traditional growing area in the wheat belt east of Perth, where more than 15,000 ha (37,000 acres) are in plantations. Currently, Western Australian sandalwood is only wild harvested and can achieve upwards of AU$16,000 per ton (Gowda and Ananthapadmanabha, 2011).

Although *S. album* is grown in small quantities, India has a virtual monopoly over sandalwood production. Over 90 per cent of India's sandalwood is grown in Karnataka and Tamil Nadu. The present Government policy pertaining to the management of sandalwood goes back to King Tipu Sultan, erstwhile ruler of Mysore. King Tipu Sultan declared sandalwood a royal tree and monopolized the sandalwood trade in 1792. Till 2002, state governments, especially Karnataka and Tamil Nadu have had monopoly control over all the sandalwood resources including those in private lands. In Kerala, however, there is no restriction on storage and transportation. This has given rise to widespread smuggling and illegal trade especially between Karnataka and Kerala. Places such as Karnataka-Kerala-Tamil Nadu border have become havens for illegal trade in sandalwood. Due to extensive illegal cutting of native Sandalwood trees, this species has become vulnerable to extinction.

There are two main commercial species of Sandalwood in the world: Australian Sandalwood (*Santalum spicatum*) and Indian Sandalwood (*Santalum album*). Annual global demand of Sandalwood was estimated to be 6,000 tons in 2006. To fulfil this demand, Western Australia exports about 2000 tons annually. The extremely high demand and reduced supply of Sandalwood is driving its price up all around the world. Sandalwood from Mysore region of southern India is generally considered to be of the highest quality available. The price of Indian Sandalwood fetches 66% higher price than Australian Sandalwood. Other species are found in the Pacific region and Australia. The natural resource of Pacific sandalwood species has been heavily exploited since the early 19th century and on some islands, the resource has been practically exhausted.

CHAPTER 2

CLASSIFICATION, GENETICS, TREE IMPROVEMENT AND BREEDING OF SANDALWOOD

2.1. Classification of Sandalwood

2.1.1 Botanical classification of Sandalwood

Kingdom Plantae – Plants

  Subkingdom Tracheobionta – Vascular plants

    Superdivision Spermatophyta – Seed plants

      Division Magnoliophyta – Flowering plants

        Class Magnoliopsida – Dicotyledons

          Subclass - Rosidae

            Order - Santalales

              Family Santalaceae – Sandalwood family

                Genus *Santalum* L. – sandalwood

2.1.2 Commercial classification of Sandalwood sold in Indian market

The classification is enumerated in to 18 classes as follows.

1. First class billet. (Vilayat budh) 2. Second class billet (China budh) 3. Third class billet (Panjam) 4. Ghotla (billets of short length) 5. Ghat badla 6. Begardad 7. Roots (First class) 8. Roots(Secondclass) 9. Roots (Third class) 10. Jajpokal (First class) 11. Jajpokal (Second class) 12. Ain bagar 13. Cheria 14. Ain chilta 15. Hatrichilta 16. Milva chilta 17. Basola bukni 18. Saw dust powder.

Sl. no.	Class	Description
1	Vilayat Budh (Class I billets)	Sound billet weighing not less than 9 kg and not exceeding 112 pieces per tonne.
2	China Budh (Class II billets)	Slightly inferior billet weighing less than 4.50 kg and not exceeding 224 pieces per tonne.
3	Panjam (Class III billets)	Billets having small knots, cracks and hollows weighing not less than 2.2 kg and not exceeding 448 pieces per tonne.
4	Ghotla (billets of short length)	Includes short and sound pieces. There are no limits of weights and numbers per tonne.
5	Ghatbadla	Billets with knots, cracks, small hollows, weighing not less than 4.5 kg and not exceeding 250 pieces per tonne.
6	Bagardad	Consists of solid pieces without limit as regards dimensions, weight or number.
7	Roots (Class I)	Pieces weighing not less than 6.75 kg and not exceeding 150 pieces per tonne.
8	Roots (Class II)	Consists of pieces weighing not less than 2.25 kg and not exceeding 448 pieces per tonne.
9	Roots (Class III)	Consists of small and side roots below 2.25 kg in weight
10	Jajpokal or Badla (Class I)	Consists of hollow pieces weighing not less than 3.10 kg and not exceeding 320 pieces per tonne.
11	Jajpokal (Class II)	Hollow pieces weighing not less than 1.3 kg per tonne.
12	Ainbagar	Consists of solid, cracked and hollow pieces weighing not less than 450 g.
13	China Sali or Large Chilta	Consists of pieces and chips of heartwood weighing not less than 2.25 g.
14	Ain Chilta	Consists of small pieces of heartwood.
15	Hatri Chilta	Consists of heartwood and chips obtained by plaining billets with Hatri or Randha (plane).
16	Milva Chilta	Consists of pieces and chips having fair proportions of heartwood and sapwood.
17	Basola Bukni	Consists of small heartwood and sapwood chips.
18	Saw dust	Sawn powder obtained while sawing the sandalwood.

(Classification of grades of Sandalwood as per Karnataka Forest Manual Rule No 95)

2.2 Genetics, tree improvement and breeding of Sandalwood

2.2.1 Genetics of Sandalwood

Genetic diversity is a prerequisite for long-term survival and adaptability of any species. It helps us to exploit the immense wealth of genes present in the nature (Young *et al.*,2000; Bahuguna, 2007). Irresponsible exploitation of this genetic wealth results in genetic erosion followed by extinction of species (Kemp *et al.*, 1993). Genetic diversity in Sandalwood is under threat owing to the illegal felling and heavy infestation by spike disease (Arun Kumar *et al.*, 2016) and therefore International Union for Conservation of Nature has categorised it as "Vulnerable".

A number of studies have been conducted to understand the genetic diversity of sandal using random amplification of polymorphic DNA (RAPD) (Shashidhara *et al.* 2003; Suma and Balasundaran 2004; Azeez *et al.* 2009) and restriction fragment length polymorphism (RFLP) (Byrne *et al.* 2003). Due to unavailability of native SSR marker, most of the diversity analysis studies in sandalwood were conducted independently with RAPD, ISSR and cross-species transferable SSR markers. RAPD analysis is an efficient marker technology for estimating genetic diversity and relatedness, thereby enabling the formulation of appropriate strategies for conservation, germplasm management, and selection of diverse parents for sandalwood improvement programmes.

Nageswara Ro (2007) developed baseline data on the levels of genetic diversity and its distribution on a geographic scale for sandalwood in India. Based on Geographic Information System, distribution map of sandal was developed and found that the populations were geographically more concentrated in the Deccan plateau. It was found that, for past 53 years, there was a monotonic decrease in the extraction of sandalwood in the state of Karnataka, India. Using allozyme markers, the genetic diversity of 19 sandal populations in peninsular India were determined and over all observed heterzygosity (H_o) was found to be 31%. The percent H_o was positively correlated with the density of the sandalwood populations ($r = 0.44$) and with increasing longitude it was found to be negatively correlated ($r = -0.51$). The dendrogram analysis indicated a clear clustering of sandal populations based on their geographic occurrence. The Deccan plateau populations were found to be genetically the most diverse and seemed to represent the 'hot-spot' of sandal genetic resources in peninsular India. These results have important implications for the conservation strategies for sandal populations in peninsular India and can be applied for the conservation of other taxa as well. Indriokoa and Rataningrum (2015) studied the impact of habitat loss caused clonality, genetic diversity reduction and reproductive failure in *Santalum album* (Santalaceae), an endangered species of Indonesia. They used the technique of allozyme and embryology observation to determine the effect of habitat loss on sandalwood grown in two of Gunungkidul populations; one was well managed as *ex situ* conservation and another was heavily wild-harvested. Result indicated significant reduction on genetic variability in harvested population may occur as result of clonality on fragmented or isolated habitat. A low genetic variability, in turn, resulted to inbreeding depression and sexual reproductive failure. Rare and missing alleles found in harvested population indicated that several alleles were not inherited to the next generation. They suggested that Clonality, biodiversity reduction and sexual reproductive failure are the main problems that should be unravelled on designing the conservation strategy of sandalwood in Indonesia.

Nurchahyani *et al* (2017), Studied the effects of population size on genetic parameters and mating system of sandalwood in Gunung Sewu, Indonesia. They found out that Genetic depletion occurred due to (i) clonality events as result of heavy-exploitation and/or natural disturbance which induced root suckering, (ii) genetic drifts and bottleneck effects, (iii) the founder effects due to parental low diversity, and (iv) the alteration on mating systems to be more inbreeders. They concluded that genetic diversity and mating systems are not much affected by population size, but more by the parental heterozygosity and the degree of clonality. Their work emphasizes the importance genetic base of populations or parental genetic diversity for maintaining an effective breeding population.

Arun Kumar *et al* (2018) studied the Allozyme variations to measure genetic diversity in Clonal Accessions of Indian Sandalwood

(*Santalum album*) and found out that presence of isozymes reflects the genetic complexity existing in the clones. They however did not find any pattern of geographic distribution of isozymes. They opined that the diversity observed among these clones would be useful in identifying diverse cross-combinations for deriving hybrids and establishing trials for production of seeds.

Ghosh *et al* (2019) derived a draft genome of *Santalum album* L. that could provide genomic resources for accelerated trait improvement. They found out a total of 37,500 genes were identified out of which 30 genes belonged to terpene synthase family (vital for oil production in Sandalwood).

2.2.2 Genetics and Karyotype analysis of *Santalum*

Zhang *et al* (2010) reported the somatic chromosome number of *S.album* was noted as 2n=20 and with basic chromosome number x =10122-124 Karyotypic analysis indicates that *S. album* was a more primitive taxon (Paliwal, 1956). In the tetraploid (2n=40) accessions analyzed, the chromosomes were relatively larger in size than those of the diploids (2n=20). This was consistent with a report by Srimathi and Sreenivasaya (1962) in which there was a 2 to 5-fold increase in the size of the chromosomes of the haustoria. There were two groups of sub metacentric chromosomes in the tetraploids; however, only one such pair was found in diploids.

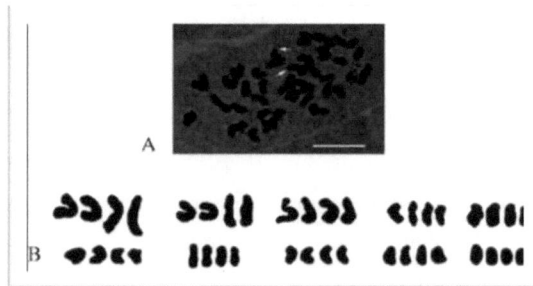

Source: Zhang et al 2010

The genus *Santalum* distributed throughout India, Indonesia, and Australia is diploid Hawaiian clades are tetraploid. By studying data from flow cytometric analysis of Pacific island sandalwoods (Zhang *et al* 2010) reported that *Santalum* appeared to include tetraploidies, ranging from diploid (n=10) to octoploid (n =40). The majority were diploid and tetraploid, and *S. album* was diploid among all the accessions investigated. Even the ploidy levels of taxa within the same section in the genus were not completely consistent. For example, *S. freycinetianum* was related to *S. album* and they were all clustered into the same section, even though *S. freycinetianum* is tetraploid. There is evidence that *Santalum* polyploids have already colonized the Pacific islands.

2.3 Tree Improvement and Breeding of Sandalwood

The implementation of a successful breeding programme for any sandalwood species will depend upon knowledge of its breeding system and its cross-compatibility with related species that are a source for potentially useful characters. Knowledge of the breeding systems of any species will help in developing strategies aimed at conserving current wild populations and establishing new plantations. Information on the breeding system and patterns of gene flow are vital for planning germplasm collection, designing and managing seed orchards and for maintaining genetic diversity in breeding populations (Zobel and Talbert, (1984). The aim is to assess the rate of gain and to compare performances between genotypes for operational use or for infusion into the next breeding program. To ensure that their performance is as consistent as possible over time and site, only half-sib families which are performing well over different environments should be selected (Namkoong *et al* 2012). It has been described that the breeding system of *Santalum* species in general is facultatively allogamous with variation found between families and individuals at the level of self incompatibility and having no ability for apomixis or parthenocarpy (Ma *et al.*, 2006; Muir *et al.*, 2007; Tamla *et al.*, 2012, Page *et al.*, 2012). This nature of outcrossing and the ability to self fertilize has provided an advantage for *Santalum* species to grow and survive in new areas (Silva *et al.*, 2015).

A general strategy for breeding program of any species (can be applied to Sandalwood too) is explained in the below section

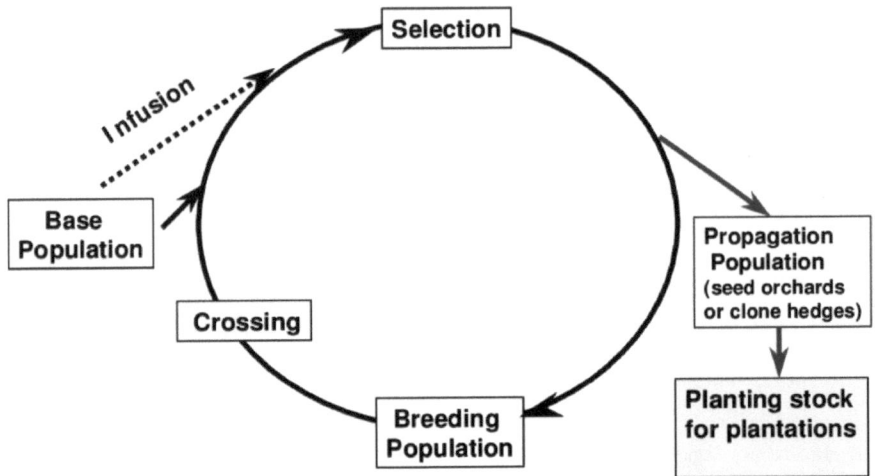

Source Boardman et al (2002)

Base population: The population of trees from which the tree breeder chooses to make selections for the next generation of breeding. In case of Sandalwood the base population can be the newly established plantations as the natural populations with adequate size are hardly available. The plantations that have been established have to be treated as base populations (first generation) for further improvement programs.

Selection: Selection is a breeding procedure wherein populations/individuals are selected based on their superior quality for desired traits. In case of Sandalwood the first phase of selection would involve "fast growth" and "disease resistance" as the desired traits. Subsequently, as the plantation grows older "more heartwood and oil content" would be the target traits. Finally, the quality of the oil produced has to be worked upon.

Propagating population: In case of perennial tree crops like sandal, due to long gestation period, the choice of propagating population has to be carried out in phases slowly adding the desirable traits as the base population ages and advanced characters show up.

Phase I - Selection and use of planting material from mother trees on the basis of fast growth and disease resistance

Phase II - Selection and use of planting material from mother trees on the basis of heartwood and oil content

Phase III- Selection and use of planting material from mother trees on the basis high content of alpha-santalol and beta-santalol.

Breeding population: The primary objective of a breeding population is to increase the frequency of desirable alleles found in the base population. While breeders know the traits they wish to improve, they do not know which alleles (genes) favourably impact the traits or their distribution in the base population. Three phases of selection described above would provide us with individuals with desired traits. Those individuals can be used for controlled crossing and improving the genetic gain over base populations. This being an advanced step in improvement would require a thorough knowledge of the trait expression and reproductive biology of the species.

Infusion: This is a process by which new superior genotypes are added from other populations elsewhere and tested for performance in the current improvement program. Such genotypes, if found successful, are added to the production population. Upon screening they are subsequently used for improvement programmes.

CHAPTER 3

SILVICULTURE AND PROPAGATION

3.1 Climate and soil

Sandal is capable of growing on all type of soils at elevation from sea level to 1800 m in rainfall of 500 - 3000 mm (Troup, 1921; Venkatesan, 1980). The formation of heartwood is said to be best in drier region at altitudes between 600 - 900 m, with moderate temperatures and rainfall of 800 - 1000 mm spread over several months. In early stages, it grows well under partial shade, but at the middle and later stages, it shows intolerance for heavy over-head shade. The tree flourishes well in different types of soil like sand, clay, red soil, lateritic loam and even in black cotton soils. It adapts well in different climatic conditions with an exception of water logged areas and very cold places (Singh and Shankar, 2007). Rai (1990) observed that it thrives best under rainfall conditions of 500 - 2000 mm and at elevations of 650 - 1200 m. The annual temperatures within the range of $10^0C - 40^0C$ are reported to be most suitable for its growth and heartwood formation. However, it can occur beyond these ranges too, but under high rainfall conditions the growth is luxuriant yet the heart-wood formation is absent or negligible. Despite its potential to grow in a wide range of conditions, plantation site needs to be carefully chosen with plenty of sunlight. Waterlogged soils should be avoided, while rich, fairly moist, fertile, iron-rich clay soils give best growth (Neil, 1990).To sum up the above reports, Sandalwood is a highly versatile plant which grows in variety of soil and climatic conditions. While selecting the site for commercial plantation as a thumb rule one should remember that sites with WATER LOGGING CONDITIONS AND SNOWFALL occurrence should be avoided.

3.2 Propagation methods

3.2.1.1 Earlier work done on seed propagation

Seed germination was given a priority among the Silviculture studies by many researchers around the world. The reason of this was due to the prolonging dormancy period and low germination rates of sandalwood seeds. These inferences were based on the facts that seeds required (1) warm stratification for dormancy loss (Srimathi & Rao, 1969) and (2) a relatively long period to germinate (Beniwal & Singh, 1989). Embryos inside *S. album* seeds are minute (Rangaswamy & Rao, 1963), as are those of other species in Santalaceae (Martin, 1946, Annapurna *et al* 2005 and Doran J, 2012,) It has been reported that propagation of *S. album* is difficult since germination is usually poor and highly variable and the germination time exceeds 12 months (Annapurna *et al.*, 2005; Singh *et al.*, 2013).

Baskin and Baskin (1998) inferred that the seeds of this species have physiological dormancy (PD) or perhaps morpho-physiological dormancy (MPD), i.e. seeds have a minute embryo that must elongate inside the seed before, during or after the loss of PD. However, Clarke and Doran (2012) have stated that no exact cause has been identified for the dormancy of *S.album* and speculated that the seeds of this species might have an exogenous kind of dormancy.

Additional information supports that the seeds of *Santalum* species are dormant at maturity and have a physiological component to their dormancy: germination increases with (1) Gibberellic acid (GA3) (Nagaveni & Srimathi, 1980; Ananthapadmanabha *et al.*, 1988; Hirano, 1990; Loveys & Jusaitis, 1994; Cromer & Woodall, 2007; Nikam & Barmukh, 2009; Gamage *et al.*, 2010) and (2) with removal of the fruit wall since the embryo has a low growth potential (Sahai

& Shivanna, 1984; Loveys & Jusaitis, 1994; Woodall, 2004; Cromer & Woodall, 2007). In contrast, Prasetyaningtyas (2007) reported that the fruit pulp apparently contains inhibitors but the extracted clean seed has no known dormancy. Dileepa *et al* (2015) studied the dormancy in *Santalum album* in Srilanka and confirmed presence of morpho-physiological dormancy associated with this species.

3.2.1.2 Collection and processing of seeds

Sandalwood occurs in groups and therefore adjacent groups in one locality should be considered as one source. In the selected seed source(s), select seed trees from which to collect seeds. In selecting seed trees, choose healthy and vigorously growing individuals. Avoid deformed, unhealthy, infected and young or over-mature trees. A seed source should ideally have more than 20 seed trees. Seeds are available in two seasons, April to May and September to October; both the seeds perform alike with respect to germination. Fresh fruits collected are depulped and dried seeds are kept for two months due to dormancy before sowing. If the seeds are soaked in 0.05% Gibberellic acid for 16 hours prior to sowing, well and uniform germination could be achieved (Rai *et al* 1990, Annapurna *et al* 2005).

Selection of mother tree: Selecting the mother tree from which the seeds are collected is one of the most crucial step in successful plantations and subsequent yield from it. Important points to be kept in mind while selecting mother tree are as follows

1. Mother tree should be at least 10 years old as it would be the right stage to assess its growth, heartwood and oil induction parameters.

2. The mother tree should be growing healthy without visible signs of disease or pests.

3. Evaluation of candidate mother tree in terms of its growth, heartwood and oil parameters has to be done in order to designate it as mother tree.

Sandalwood seeds when collected from the mother trees are rinsed in clean water and then immersed in water (floatation) to isolate empty seeds from filled seeds. The clean seeds are then immersed in water. The seeds that sink give the best grade while any floating seeds can be retained as lower grade seed or may be discarded (Ananthapadmanabha, 1988). It is important to use good-quality seeds because they produce strong, healthy seedlings. Seeds germinate best when they are sown in a free-draining medium such as pure sand or 2:1 mix of river sand and soil. Composted sawdust is also a very good germination medium. (Page, *et al* 2012)

Various stages of seed processing and seed germination Vietnam Sandalwood Group nursery Ha Noi.

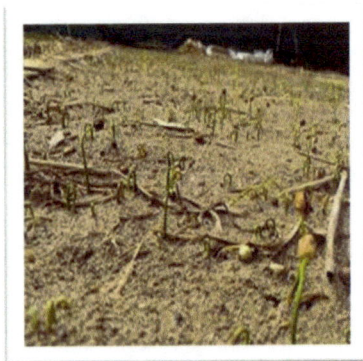

20 days after sowing in bed   *30 days after sowing in bed*

40 days after sowing in bed

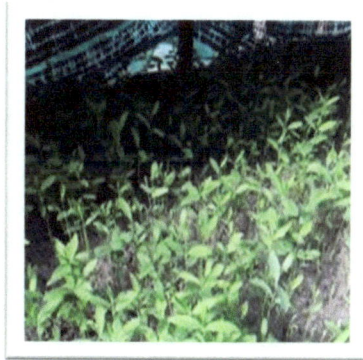

50 days after sowing in bed

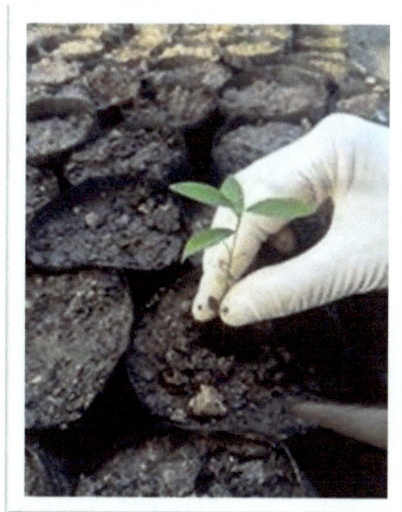

Transplanting the seedlings in polythene bags (2-4 leaf stage)

Flow chart of the seed collection and processing of Sandalwood seeds

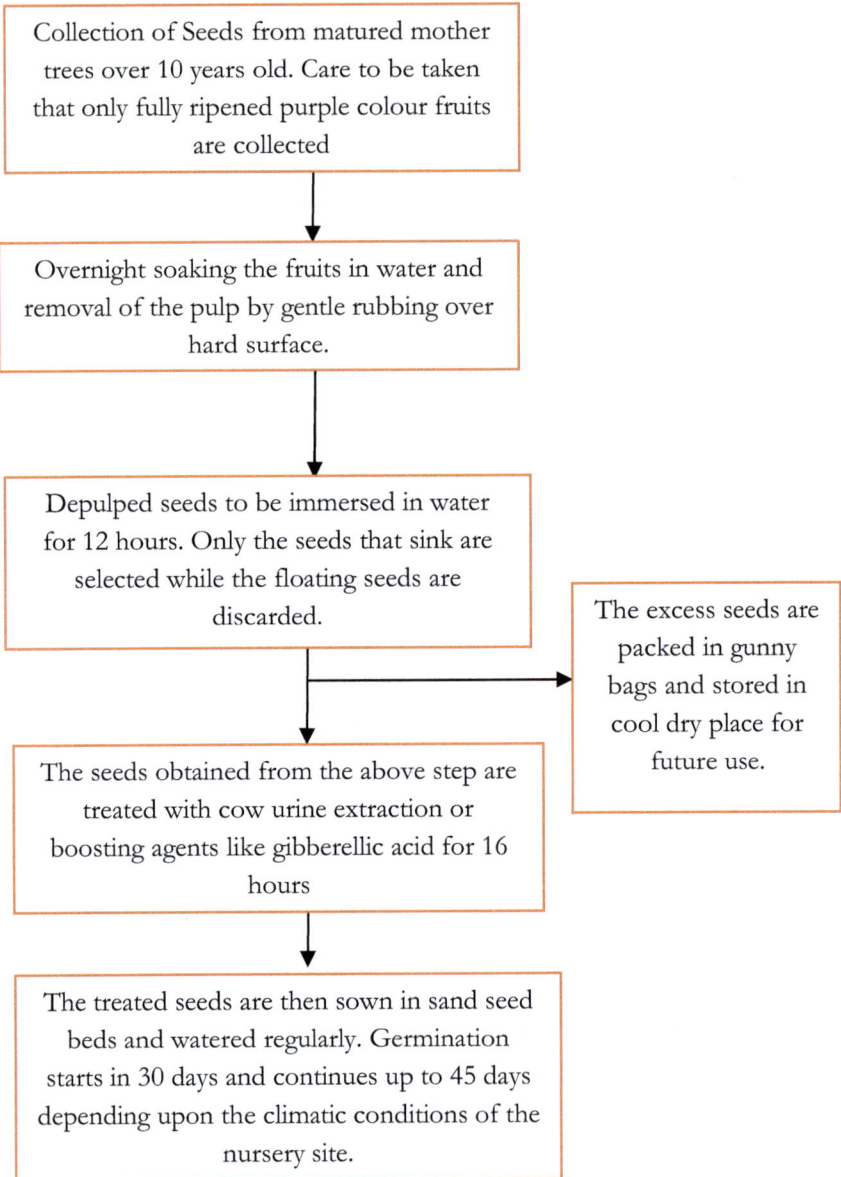

Collection of Seeds from matured mother trees over 10 years old. Care to be taken that only fully ripened purple colour fruits are collected

↓

Overnight soaking the fruits in water and removal of the pulp by gentle rubbing over hard surface.

↓

Depulped seeds to be immersed in water for 12 hours. Only the seeds that sink are selected while the floating seeds are discarded.

→ The excess seeds are packed in gunny bags and stored in cool dry place for future use.

↓

The seeds obtained from the above step are treated with cow urine extraction or boosting agents like gibberellic acid for 16 hours

↓

The treated seeds are then sown in sand seed beds and watered regularly. Germination starts in 30 days and continues up to 45 days depending upon the climatic conditions of the nursery site.

3.2.1.3 Seed bed preparation, sowing and maintenance:

In order to ensure favourable germinating conditions for the seed, the seed bed has to be designed in such a way that it provides sufficient water retention, ample root growth space and proper drainage. In order to achieve that, following structure of the seed bed may be followed:

Brick boundary: A two brick mounted boundary has to be made to ensure dimensional stability of the seed bed. The width of the seed bed has to be 3-4 feet while the length can vary as per the space available in the nursery but not more than 20feet.

Base layer: After the brick boundary has been prepared, the first layer of small stones up to a thickness of 3 inches has to be laid. This would ensure proper drainage of water allowing free breathing and growing space for the seedlings.

Pebble layer: Over the stone layer another layer of clean and washed pebbles has to be laid up to 3 inches thickness. The objective of this layer too is to provide proper drainage and ample root growth space for seedlings.

Sand layer: The third layer of the seed bed is of river sand, the river sand to be used should be properly sieved and washed before being used to avoid any contamination to the young seedlings. The thickness of this layer should be about 4 inches.

Seed layer: The fourth layer of the bed is the seed layer. The treated seeds have to be uniformly spread over the bed and gently tapped so that they are firmly embedded in the sand below.

Top sand layer: Once the seed are spread over the sand layer and gently tapped, a thin layer of ½ inch has to be spread over the seeds so that they are covered completely with sand. Once the seeds are covered completely with sand, gentle watering has to be done using rose can. This would show the patches where the seeds are exposed and are not completely covered. Such patches should be covered by spraying extra sand so as to cover the seeds completely.

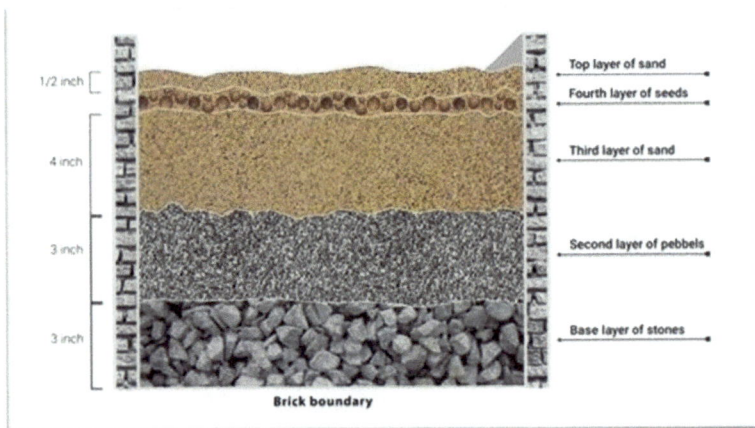

Seed bed design for sandalwood germination

Seed bed maintenance: Once the seeds are sown, water the bed with a solution of 2g/litre solution of Mancozeb (Non systemic broad spectrum fungicide) to prevent attack of fungus to the germinating seeds and emerging seedlings. Application of this fungicide has to be done every 15 days throughout the seed germination and seedling development cycle in the seed bed.

3.2.1.4 Transplanting of the seedlings:

Step 1: The seeds start germinating in 30 days and are ready for transplanting by 40-45 days after sowing. By this period the seedlings reach two-leaf stage which is the most ideal stage for seedling transplanting to polythene bags. Transplanting should be carried out during early morning (7-10 am). The seed bed should be thoroughly watered an evening before the transplanting is planned.

Step 2: The sand of the bed has to be loosened gently using a hand held shovel and the seedlings extracted without damaging the root system of the seedlings. The extracted seedlings should be dipped in a bucket containing 2g/litre of Mancozeb for surface cleaning and to avoid drying of the seedlings.

Step 3: Care must be taken that half-filled bags (4x6 inches size) of potting mixture are ready before hand in order to ensure timely transplanting. Ideally, polybags should contain soil mixture of ratio 2:1:1 (Sand: Red Soil: FYM). The ratio however, can change depending upon the nature and structure of individual components. This has to be standardised for each nursery depending upon the local conditions.

Step 4: The seedlings extracted from the seed bed for transplanting have to be kept in shade and the transplanting carried out during early hours of the day. Care must be taken that only healthy seedlings with lush green growing tip and well developed roots are only used for transplanting. Crooked, malformed and diseased seedlings have to be discarded at the transplanting stage only to avoid mortality and wastage of time, money and efforts.

Caution!!!! It may be noted that over grown seedlings in the nursery bed (6 leaves to 8 leaves) result in over 30-40% mortality of the transplanted seedlings. The seedlings that survive also shed the older leaves first and then start growing with new leaves. This results in slowing down of the nursery schedule for over a month. Therefore meticulous planning has to be done in advance for successful nursery production.

3.2.1.5 Nursery hosts of Sandalwood

The plants with which sandalwood forms haustoria are called hosts. Sandalwood forms haustoria with many different species, but some species (particularly legumes) support greater growth and vigour in the sandalwood. There are three main host types used for cultivating sandalwood (Page *et al* 2012)

• Pot host—planted in the poly bag after the seedling reaches the 2-4-leaf stage and typically persists in the field for a time after planting

• Intermediate host—small tree or large shrub, typically a short-lived (about 5 years) nitrogen-fixing legume that is planted close to the sandalwood

• Long-term host—large tree that provides a host for the entire sandalwood rotation; it is planted at a lower density in the plantation and at least 3 m from the closest sandalwood tree.

Nursery Hosts

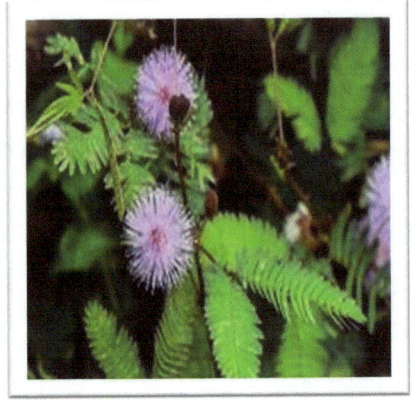

Alternanthera nana/ Mimosa pudica

The pot host should be planted as stem cuttings after the sandalwood is at the 4–6-leaf stage. If the pot host is planted too early, it will grow quickly to overtop the young seedling, leading to stunted growth and might even result in death of the sandalwood. However, if the pot host is planted too late, the sandalwood may grow slowly. It may be noted that the pot host has to be regularly pruned to avoid completion for the sandalwood.

3.2.1.6 Maintenance of seedlings

The freshly transplanted seedlings have to be then kept in a shade house (50% sunlight penetration). The seedlings have to be regularly monitored and preventive sprays of mancozeb have to be carried out every 15 days to avoid any fungal attack. In case of any infestation of fungus/ insects is seen in the nursery, the affected seedlings have to be immediately quarantined and removed from the main lot of seedlings. This would help in avoiding the spread of infestation to the other healthy seedlings. Once the seedlings reach the 10–12-leaf stage, they need to be progressively moved to areas of higher light levels to 'harden' or acclimatise them to conditions in the field. They should have 50% shade in the first month, and then 25% shade for another month, and then one to several months in full sun, before they are planted in the field. Seedlings transplanted directly from the shade to a full-sun position in the field typically have poor survival.

1 month old seedlings

2 months old seedlings

3 months old seedlings

4 months old seedlings

5 months old seedlings

6 months old seedlings

Seedling in Vietnam ready for transplantation in the field

A seedling is ready to plant in the field when it shows at least two of the following signs:

- The seedling is actively growing, with new shoots, and has deep green leaves.

- The bottom of the stem is slightly woody (i.e. changes from green to brown).

- Height is about 25–30 cm.

- Have at least 10 pairs of leaves

- The seedling has small branches developing at the junction of the leaf and main stem.

- Some small roots are visible through the holes in the bottom of the plastic.

3.2.2 VEGETATIVE PROPAGATION

3.2.2.1 Marcotting

Marcotting is a process in which roots are induced to develop on a branch (stem) while it remains attached to and nutritionally supported by the parent plant. Once the roots are developed, the branch is detached and becomes a new individual growing on its own roots and is referred to as a marcot or a layer. Rooting in marcotting is initiated by girdling (removing away a section of bark about 3 cm wide around the circumference of the branch) the branch thus interrupting the downward flow of photosynthates through the phloem from An air-layered branch with roots the branch tip. Efforts for marcotting (Air-layering) have not being reported successfully in *Santalum album*. However, Mwa'gingo *et al* (2006) reported successful air-layering in *Osyris lanceolata* (African Sandalwood). They reported that observed that rooting success of up to 80% can be achieved from air layers, making this propagation technique a viable alternative to seedlings or cutting propagation. Rooting success was influenced by both the season and application of rooting hormone with optimal rooting being achieved during June and September with the addition of IBA at a rate of 50 ppm.

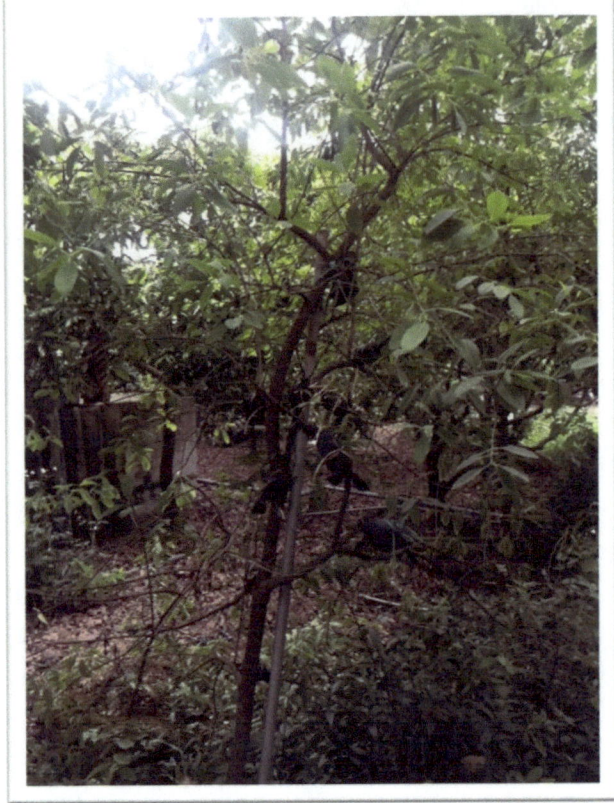

Sandalwood marcotting in Quoc Oai district, Viet Nam

3.2.2.2 Grafting

Clonal propagation of sandal is achieved by cleft grafting on seedlings at least one year old, and by propagating root-suckers. However, propagation through shoot-tips of branch-cuttings have not been successful. Cleft grafting has been carried out successful for developing Clonal Seed Orchards of Sandalwood by Institute of Wood Science and Technology, Bangalore a pioneering research institute of Sandalwood under Indian Council of Forestry Research and Education , Dehradun (Anon, 1992)

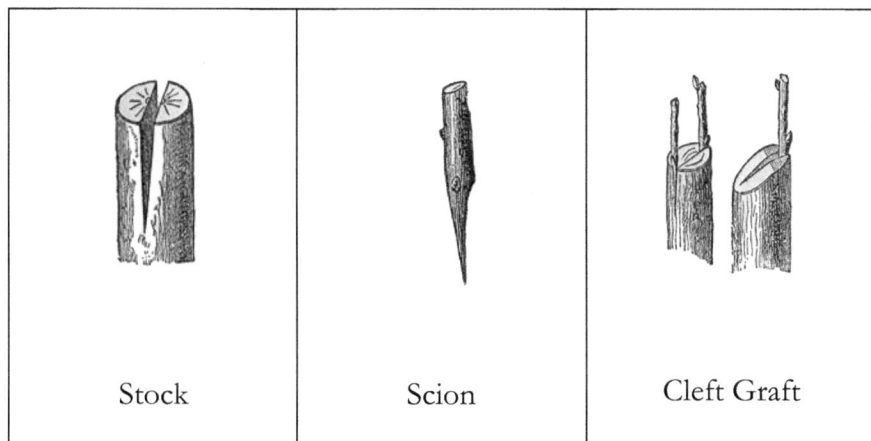

| Stock | Scion | Cleft Graft |

3.2.2.3 Tissue culture

Yeast extract was found to initiate and stimulate the proliferation of sandal endosperm. However, for a satisfactory proliferation an auxin (2-4 D) and a kinin (Kinetin) in conjunction with yeast extract are required (Rangaswamy and Rao, 1963). Sandal plants have been successfully micropropagated by *in-vitro* methods either using juvenile or mature plants (Lakshrni Sita 1986; Rao and Bapat 1992).

Rathore *et al* (2007) described micro propagation methods through axillary shoot proliferation have high potential for rapid and mass production of Clonal planting material of *S. album* from plus trees. Rathore *et al* (2014) studied the effect of sucrose, agar-agar concentration and pH of the media on somatic embryo induction, maturation and germination from the explants of the mature trees and found that during maturation, size of the somatic embryos increased with increase in the concentration of sucrose in the medium favouring adventitious shoot induction and embryo creaking.

Source: Cheng et al 2017

Although there are various reports of developing tissue culture raised plants for sandalwood. Any success story of raising plantation from this method and harvesting a crop with high heartwood and oil content is not available. A more cautious approach has to be adopted for tissue culture method with sufficient genetic base of desirable traits used as the mother source of tissues. Maintaining of sufficiently large genetic base is of prime importance in case of *Santalum album* tissue culture because of long gestation period and possibilities of its exposure to wide range of insects and pests during its life time.

CHAPTER 4

SANDALWOOD NURSERY

ESTABLISHMENT AND MANAGEMENT

A nursery is a managed site, designed to produce seedlings grown under favourable conditions until they are ready for planting. All nurseries primarily aim to produce sufficient quantities of high quality seedlings to satisfy the needs of users (Ratha Krishnan *et al* 2014). Nurseries play an important role in successful plantations programs as they are the first step in the whole plantation process. Nursery is a place where the future plantations are nurtured, therefore it is very important for any successful plantation program to strategically set up and manage the nurseries.

4.1 Factors to be considered before establishment of nursery

- The nursery site should be located in the nutrient rich/medium soil, near to water source, free from soil pathogens and insects.

- Availability of cheap and skilled labours

- Good access to the main road for easy transportation so that delivery of nursery materials and transportation of seedlings to the planting sites is easy and safe.

- Power Supply in forest nursery site should be uninterrupted because electric power is needed for Lighting and the operation of nursery equipment, including irrigation pumps, refrigerator and so on. If electricity supply is not available, then diesel-powered generators will have to be used.

- Forest nursery site should be close to natural water sources, such as rivers, waterfalls, streams, Lakes, ponds and wells. The water should be clean and free from pollutants and industrial wastes that might be detrimental to plant growth

- The site should be on gently sloping area and away from other tall crops: this is important for good surface drainage as well as to encourage air circulation.

- An appropriate site must be selected for the most effective, efficient, and economical design of a nursery. The purpose and target of plants to be produced will decide the site selection and its improvement.

- Careful observation of site conditions and an assessment of past and present climatic records are important for healthy growth of the plant species intended to be grown.

4.2 Design and layout of the nursery

A well designed forest nursery should have proper roads, fencing, office, potting and transplanting sheds, seed germination and growing areas, water supply, telecommunications, water sprinkler system etc. Consideration should also be given to the construction of storage facilities for nursery equipment, tools, fertilizers, pesticides and other materials. Paths, trails & road system are necessary for the efficient movement of materials & personnel within the nursery. For permanent nurseries, it is recommended to install overhead water system (sprinkler) by conducting water to the nursery through the pipes & distribute it over the seedlings as a fine spray with low pressure system with a pressure of 30 psi or more.

LAYOUT OF SANDALWOOD NURSERY

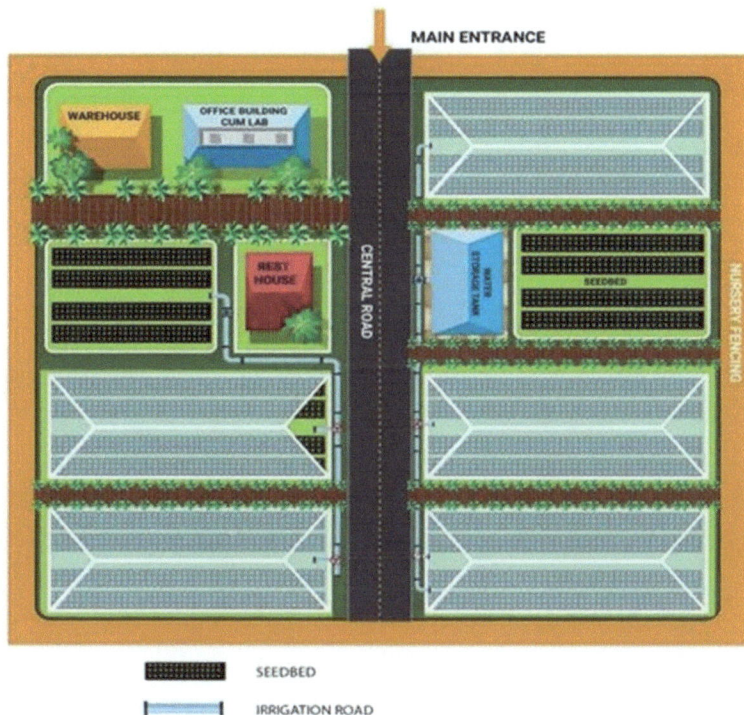

MAIN ENTRANCE

WAREHOUSE

OFFICE BUILDING CUM LAB

CENTRAL ROAD

REST HOUSE

WATER STORAGE TANK

SEEDBED

NURSERY FENCING

SEEDBED

IRRIGATION ROAD

4.3. Nursery management

In addition to appropriately managing the pot host and progressively hardening the seedlings, attention to hygiene, preparation of growing medium, drainage, and appropriate watering and fertilisation regimes will provide good results. Since sandalwood is a very sensitive species during the initial stages of its life cycle, utmost care has to be taken of it during nursery production. Few of the important aspects to be taken care of are enlisted below.

Hygiene

Good hygiene in the nursery can ensure the development of healthy seedlings. The key is to keep the nursery free from plant debris (e.g. pruning's, fallen leaves, dead seedlings), which will limit the potential hot spots for disease development. Many unexplained problems in the nursery can be attributed to diseases caused by unseen fungi and bacteria. Sterilising the growing medium can help to prevent diseases being brought in by the soil.

Fertiliser

Improved growth can be achieved by using fertile soil. A friable, rich volcanic soil will have more nutrients than a heavy clay soil. Additional nutrients can be added to the soil by applying dilute solutions of worm castings, composted organic matter, or commercial liquid and/or slow-release fertiliser. Nutritional well-being of the seedlings is of great importance because healthy seedlings perform well in the field.

Drainage

A poorly draining medium can lead to waterlogging, particularly during the wet season. Waterlogging is one of the main causes of seedling death in the nursery. It is therefore important to include sand in the medium to improve drainage. It is suggested that the sand percentage in the medium of seedling growth should be at least 50% to avoid compaction and root rot in the nursery.

Watering

The watering regime for seedlings raised in a nursery will vary according to the amount of rainfall and the position of the seedling in the nursery (e.g. 50% shade versus full sun). As a thumb rule watering should be done on the basis of requirement of the plant rather than based on a schedule. For example, on warm days when the potting mixture is dry and plants show signs of drying watering should be done every day whereas on a cloudy day when the heat is not too strong it should be done once in two days. Time of watering during the day is also important. Watering should be ideally done before 8 am in the morning or after 4 pm in the evening. Watering during peak sun heat (11 am to 3 pm) is not advisable as it may result in excessive transpiration losses and stress to the young tender plants.

CHAPTER 5

ESTABLISHMENT OF SANDALWOOD PLANTATION

5.1. Selection of site

As indicated earlier sandalwood grows on variety of soils viz, red lateritic soil, moderate black soil, loamy soils etc. It is worth to get a soil test done before the plantation is taken up. Sandalwood tree cultivation requires pH of the soil in between 6.5 to 7.5 with a little bit of alkaline. Well-drained soils which have a good organic matter are also excellent for sandalwood cultivation. Sandal trees grow mainly on red ferruginous loam, overlying metamorphic rocks, chiefly gneiss. They can tolerate shallow, rocky ground and stony or gravelly soils, voiding saline or calcareous soils, and are not exacting about the depth of the soil. Rich and moist soils such as well-drained alluvial soils do not support the heartwood and oil formation. Although the growth of the tree will be excellent in these soils the heartwood formed in such trees will be deficient in oil. Conversely, the plants grown on stony/gravelly soils yield better quality heartwood and oil.

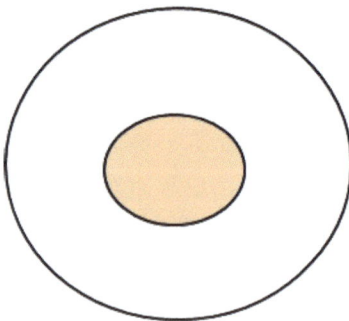

Cross section of tree in loamy soils   *Cross section of tree in stony/gravelly soils*

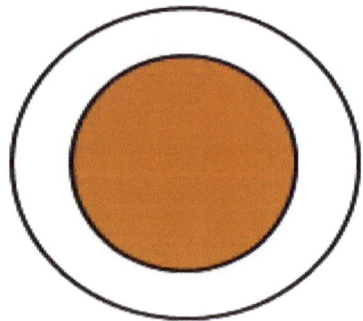

Before starting plantation of sandalwood summer ploughing is to be carried out up to a depth of 1 ft followed by a rotavator to break the clods. This soil has to be exposed to summer sun to kill any insect eggs/larvae and weed seeds (Solar sanitization). Also, prepare the soil in such a way that the water in heavy rain or floods is easily come out from the drain. Sandalwood tolerates a wide range of soils so long as they are well drained but does not tolerate areas prone to water logging. Sandalwood should not be preferably planted in areas prone to frost.

5.1.1 Biophysical considerations for site selection

The more information there is available about the site conditions in the area being considered for tree and shrub planting, the better are the chances of selecting the tree and shrub species best suited to the area. Information most commonly included in site reconnaissance is:

- *Climate* - temperature, rainfall (amount and distribution), relative humidity, and wind.

- *Soil* - depth of soil and its capacity to retain moisture, texture, structure, parent material, pH, degree of compaction, and drainage.

- *Topography* - important for its modifying effects on both climate and soil.

- *Vegetation* - composition and ecological characteristics of natural and (when present) introduced vegetation. On areas which have not been degraded by man, the vegetation can provide an indication of the site. Unfortunately, over much of the tropical and arid world, the vegetation has been so disturbed that it is no longer a reliable indicator of potential planting sites; in these situations, site selection should be based on soil surveys.

- *Other biotic factors* - past history and present land use influences on the site, including fire, domestic livestock and wild animals, insects and diseases.

- *Water table levels* - a knowledge of the depth and variation of the water table levels in the wet and dry seasons is valuable and can be crucial in determining the tree and shrub species that can be grown. Water table levels can be estimated from observations in wells or by borings made for this purpose.

- *Availability of supplementary water sources* - ponds, lakes, streams, and other water sources so that the irrigation schedule of the plants is not affected in case the primary water source fails.

- *Distance from nursery* is an important aspect as it determines the operational ease for establishment of the plantation. As mentioned in earlier section, the nurseries have to be established in the vicinity of the planting site. This would facilitate the climatic acclimatization of the seedlings during the course of nursery production. This would ensure better survival and growth of the plantations.

5.1.2 Socio-economic factors to be considered for site selection

- The availability of sufficient labour in order to carry out plantation management related activities.

- Motivation of the local population for smooth working of the plantation management plans.

- The distance of the forest plantation to the market and consumer center is one of the important factor in deciding the plantation site. Ideally, potential market for the produce of the plantation should be located within 50 kms of the plantation site. In case of sandalwood, where the host plant provided is generally a horticultural crop, it is all the more important to have a viable market in close vicinity of the plantation site.

- Sandalwood being a very high value crop it is very important that the land ownership and the lease tenure are very clearly documented. This would ensure a clear defining of the owner of the sandalwood and the certification of the produce shall be easily done.

5.2. Land preparation

Land preparation for the plantation is an important factor that decides its success. It mainly aims to remove competing vegetation from the site, create conditions that will enable the soil to catch and absorb as much rainfall as possible, provide good rooting conditions for the planting, including a sufficient volume of rootable soil and create conditions where danger from fire and pests is minimized. Site preparation is directed toward giving the seedlings a good start with rapid early growth. In general, the methods used to achieve site preparation will vary with the type of vegetation, amount and distribution of rainfall, presence or absence of impermeable layers in the soil, the need for protection from desiccating winds, and scale of the planting operations. Additionally, the value of the tree or shrub crop to be grown is important in determining the amount of expense that may be justified in plantation establishment.

5.2.1 Methods of site preparation

5.2.1.1 Soil and water conservation structures

As mentioned in the earlier section, site preparation is carried out to provide ideal growing conditions to the plant at the same time conserving soil and water in the plantation area. The kind of soil working carried out depends up on the soil type, slope of the land and rainfall in the area. Based on the above factors different types of pitting and soil working suggested are as follows (Anon, 2020):

5.2.1.1.1 Earthen bunds

Bunds are small embankment type structures made up of locally available earth materials. Land slope and soil characteristics are considered for selection of bund type and design. Bunds help to check the velocity of the run-off, to carry excessive rainfall safely downstream and to let off stream flow in natural channels. Bunding increases the time of concentration of rainwater thereby allowing rainwater to percolate into the soil. Where ever possible agronomic conservation measures like agrostology, planting of grass specials etc are provided on the constructed bunds.

Overview of a water harvesting system.

5.2.1.1.2 Stone pitched contour bunds

Stone Pitched contour bunds are constructed in contour at suitable intervals in slopes. The adoption of this intervention has led to reduction in soil erosion and increased water availability for the plants.

This type of construction is very suitable for laterite soil or wherever stone is available, up to 35%of slope areas can be protected by this way.

5.2.1.1.3 Graded bunds

Graded bunds are adopted in areas having low infiltration (< 8 mm/hr) and more than 800 mm rainfall. Graded bunds are laid along pre-determined longitudinal grade instead of along the contours for safe disposal of excess runoff. Gradient given may vary from 0.4 t0 0.8%.(0.4 for light soils and 0.8 for heavy soils).

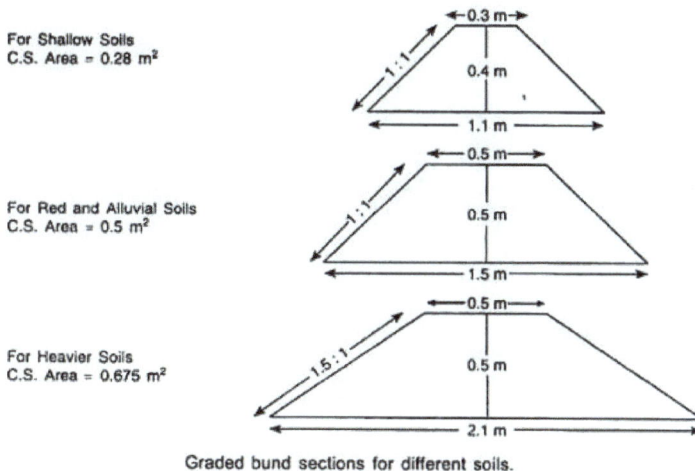

For Shallow Soils
C.S. Area = 0.28 m²

For Red and Alluvial Soils
C.S. Area = 0.5 m²

For Heavier Soils
C.S. Area = 0.675 m²

Graded bund sections for different soils.

5.2.1.1.4 Vegetative hedges

Run-off velocity can be reduced drastically by planting vegetative hedges, bunch grass, or shrubs on the contour at regular intervals .These hedges can increase the time for water to infiltrate into the soil and facilitate sedimentation and deposition of eroded material by reducing the carrying capacity of the overland flow. Vegetative hedges or narrow grass strips serve as porous filters. These hedges may not reduce runoff amount but can drastically decrease soil loss.

5.2.1.1.5 Trenches

Contour trenches are used both on hill slopes (Slope > 33%) as well as on degraded and barren waste lands for soil and moisture conservation and afforestation purposes. The trenches break the slope and reduce the velocity of surface runoff. It can be used in all slopes irrespective of rainfall conditions (i.e., in both high and low rainfall conditions), varying soil types and depths.

- *Contour trench*: Trenches are constructed as continuous across slope with 45-50 cm depth and bottom width and trapezoidal in shape

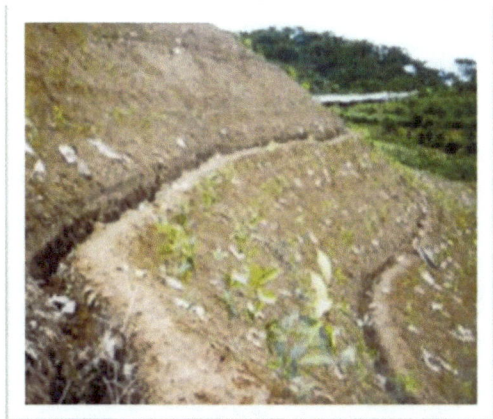

- *Staggered trench*: The length of trenches is kept short up to 2-3 m and its spacing is 5-7m. It is suited for medium rainfall areas with dissected topography.

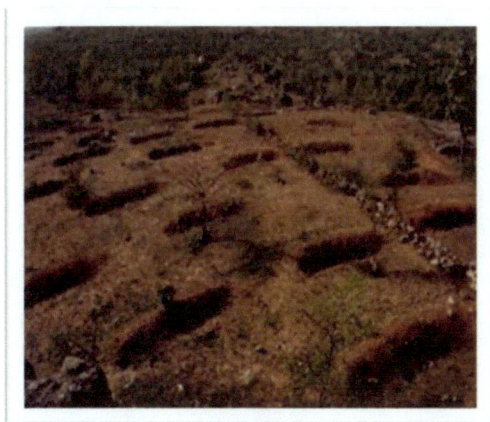

5.2.1.1.6 Strip terrace

It is used to control soil erosion in highly sloped areas. It basically involves construction of ridges and step like structures across land slope.

5.2.1.1.7 Moisture conservation pits

Any form of depression or micro pit is constructed over the land surface to arrest excess surface runoff and silting and thus leading to ground water recharge. Pits of suitable dimension are constructed in the field which would impound water and contribute to ground water recharge during rainy season.

5.2.1.1.8 Farm ponds and water harvesting structures

Farm Ponds are mainly meant for the purpose of storing the surface runoff. The farm ponds and water harvesting structures constructed in the low lying areas contributed to the conservation of excess rain water and the replenishment of ground water. The water harvesting structures are constructed as masonry structures. It will increase the soil moisture regime around the structure and recharge the ground water.

5.2.2 Design the planting garden

The pits size of 45 x 45 x 45 cm should be dug during soil/land preparation. The plant-to-plant distance should be 10 feet. Make sure there will not be any stagnated water in pits before planting. Expose pits for sun for a couple of months to dry up the pits any pests will be destroyed (Solar sanitization).

5.2.2.1 Marking and pitting

When Sandalwood is established as a plantation crop, stakes should be prepared in advance, at least a week before the onset of the rainy season. The stakes are used to mark the planting spots for pitting. During staking, a spacing of 2.5 m x 2.5 m should be adopted in the highlands while a spacing of 3.5 m x 3.5 m should be used in the semi-arid areas. Pitting should be done at least 2 weeks before the onset of the rainy season. In the highlands, Sandalwood should be planted in round pits of 30 cm deep and 30 cm diameter. In the semi arid areas, it should be planted in square pits of 45 cm x 45 cm x 45 cm.

5.2.2.2 Spacing in the sandalwood plantations

Spacing to be followed in the plantation depends upon the kind of farmer we are dealing with. In case of farmer that requires immediate yield from the host plants a wider spacing of 20x15 ft between the sandalwood plant and one host planted at 10 feet in between is suggested. This would allow ample space for the sandawood and host to grow.In case where the farmer can wait for longer period to get benefits from the plantation a closer spacing of 15 feetx 10 feet is suggested. Pits of 45X45X45 cm^3 are dug out at a spacing of 10 feet x 10 feet. Healthy sandal seedlings are planted in alternate pits at a spacing of 20 feet within a row. Host plants are planted in between two sandalwood plants in a row. This configuration would result in 200 plants of sandalwood and 200 plants of host per acre.

5.2.2.3 Grading and transportation of seedlings

Before the planting stock is taken to the site for planting, it is utmost important to carry out grading of the seedlings so that only healthy seedlings are transported. Selection of seedlings should be based on the height (20-30cm), Collar diameter (> 2mm) and brown main stem with healthy green growing tip of the seedlings. Any negligence in this step can result in the mortality of the plants in the field and loss of time and efforts. It is also suggested that the seedlings are carried to the planting site 10-15 days in advance of planting time so that the seedlings get acclimatise to the local conditions.

5.3 Planting of seedlings and hosts

Planting should be done during the onset of rainy season, preferably after first couple of showers. Extra care has to be exercised during the transportation of seedlings to the planting site to avoid any damage. The seedlings have to be carefully taken out from the poly cover without damaging the sandalwood root system and the host plant. The seedling along with the host should be placed at the center of the pit and the soil filled. Basal dose of fertilisers (1kg compost + 100 grams Single Super Phosphate + 100 grams Neem cake + 10 grams of Thimet) may be applied in the half filled pits and thoroughly mixed with the soil before the planting is done. Sandalwood grows well with hosts such as Pomegranate, *Phyllanthus emblica, Psidium guajava,* lemon, avocado etc. to increase the yield of farmer and make the plantation self-sustaining. A plant of Casuarina equisetifolia should be planted about 1.5 ft away from Sandalwood to provide long term nutritional requirement of the plant. This would also ensure that the horticultural host is protected from the ill-effects of Sandalwood in long run and its productivity will not be affected to a great extent.

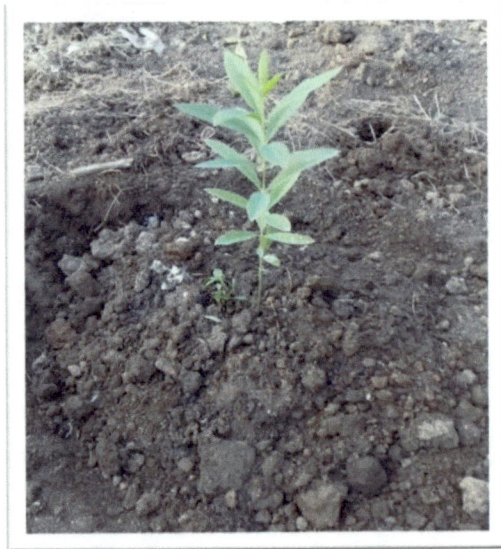

Seedling removal and planting in the field

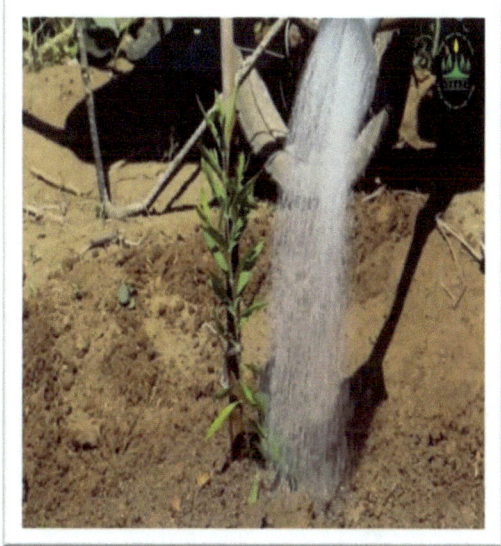

Post planting irrigation and care

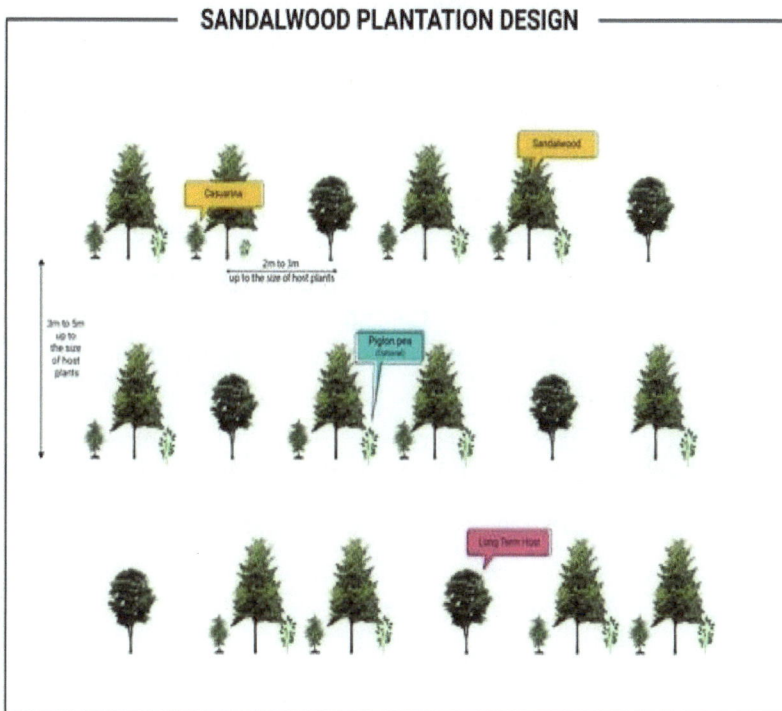

SANDALWOOD PLANTATION DESIGN

CHAPTER 6

SANDALWOOD PLANTATION MANAGEMENT

6.1 Sandalwood growth management

A plantation is the large-scale estate meant for farming that specializes in high value cash crops. These plantations are different from the natural stands in a way that they are man-made, consist of one or two species and are intensively managed. The main aim of these commercially managed plantations is to produce larger biomass of desired quality in shorter rotation period as compared to natural forests (Savill *et al* 1997).

Sandalwood plantations are mainly grown with an objective to produce more heartwood and oil from the trees per unit time. In nature, a sandalwood tree takes about 30-60 years for producing heartwood and oil of commerce. Sandalwood plantations world over have been established and are being maintained with a rotation period of 15-20 years (Gowda and Ananthpadmanabha, 2012). If we have to shorten the rotation period from 30-60 years to 15-20 years it is imperative that intensive management of the plantations have to be carried out. A growth management plan is described below which can be followed for commercial plantations of sandalwood.

GROWTH MANIPULATION PLANS FOR SANDALWOOD PLANTATION

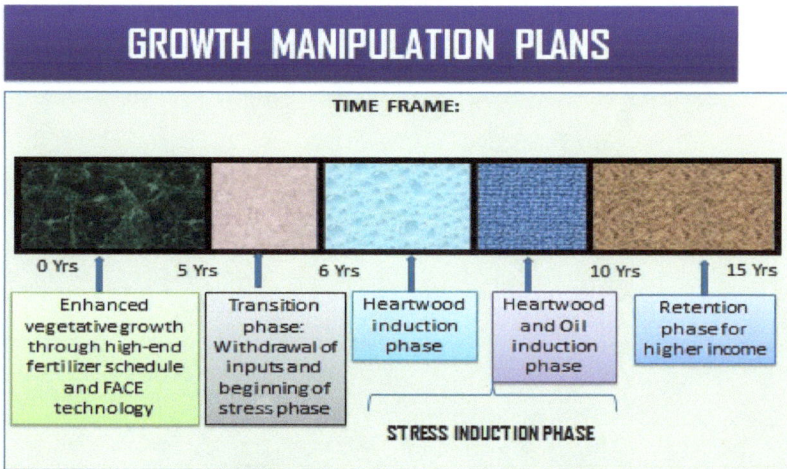

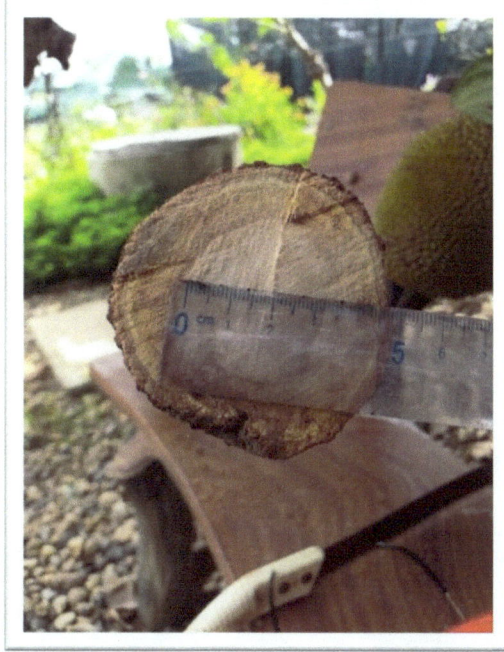

Heartwood formation in the plants of four year old in Vietnam

PHASE-I: VEGATATIVE GROWTH PHASE

In order to achieve maximum vegetative growth, for the first five years, the plantation will be intensively managed with strict irrigation and fertilizer application schedule. It is targeted to develop trees with a height of 5 meters and girth of 40 cms in these five years.

PHASE II: WITHDRAWAL

Once the trees have attained the desired height and girth, the second phase would be slow withdrawal of all the management inputs. This phase would last for three years from 6th to 9th year. It would be marked by the initiation and development of heartwood in the trees.

PHASE III: HEARTWOOD AND OIL INDUCTION

This phase will be characterized by the treatments to induce **"ARTIFICIAL STRESS"** in the plants to fasten the process of heartwood and oil induction. This phase will last for three years from 10th to 12th year. Regular monitoring of the progress of heartwood and oil induction will be carried out through core sampling and analysis.

PHASE IV: SELECTIVE HARVESTING

Selective harvesting can be carried out in the plantations where in the best grown trees @ 10% of the population would be harvested in the 13th year and 14th year. Remaining 80% of the surviving trees will be harvested in the 15th year.

As indicated in the above section, first five years of the plantation are extremely important as it is during this phase maximum biomass is produced. Accordingly, the inputs and management operations have to be of best quality and as per strict schedule.

6.2. Silvicultural operations

6.2.1 Tree shaping

Tree shaping is very important Silvicultural operation in sandalwood plantations in order to ensure that the host and sandalwood don't compete for nutrients and light. The aim of this operation is to see that the sandalwood grows like a conical crown tree while the host is grown like a bush. This can be achieved by removal of all the branches that are growing upward and that could take over the leader. The aim is a straight and clear bole 2-4 metres long (depending on the fertility of the soil). Therefore, it is recommended that pruning begins in the second year and is repeated regularly every two years until the stem has reached the targeted length.

Mature trees that have been shaped properly in their youth seldom need trimming to shape, but when trees have lost branches in storms or grown faster on one side due to root damage, the remaining branches can be reduced to re-shape the tree.

6.2.1.1 Formative pruning

Formative pruning (also referred to as framework pruning) is carried out in the first three years of planting a young tree to create or 'form' the shape, and establish a framework of main branches. This type of pruning is done in the initial stage (between 24-36 months age) of the plantation with an objective of allowing the central leader of the plant to grow and suppress the side branches competing with the central leader. This step ensures rapid height growth of the plant with healthy growing tip. This is first step to achieve the form of tree plantation manager intends to have. ***Note: Never remove more than one-fourth of a tree's branches at one time. Remember: it is better to make several small pruning cuts than one big cut. Avoid cutting large branches when possible.***

Framework Pruning Central Leader Tree Form

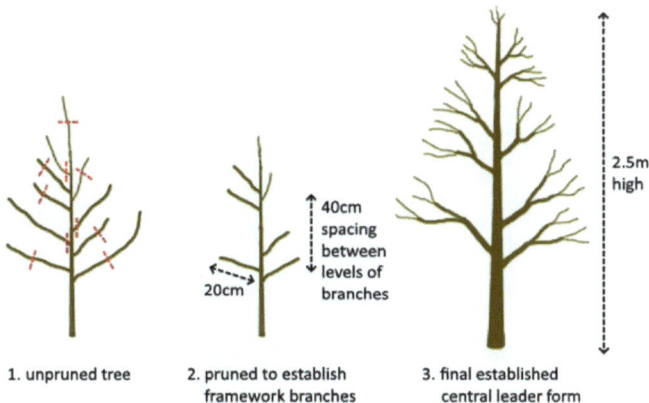

1. unpruned tree
2. pruned to establish framework branches

40cm spacing between levels of branches

20cm

3. final established central leader form

2.5m high

Courtesy: Angelo Eliades and Deep Green Permaculture, Australia

6.2.1.2 Form pruning

Often a tree is not pruned for a year or more and needs to be pruned to bring it back to being a tree with a single trunk. Form pruning is distinct from formative pruning in that it requires knife or secateurs. This method is effective for young saplings up to 4 years old, it involves the identification of forked branch and its removal using a knife or secateurs also this type of pruning involves removal of any branches growing vertically towards the main central leader and competing with it.

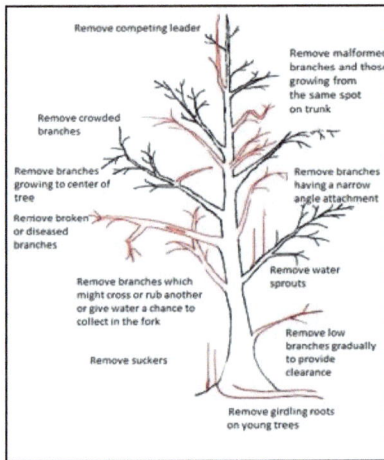

Source: Edward F. Gilman and Amanda Bisson (2019)

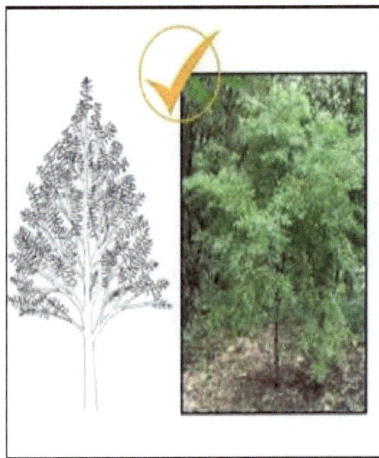

Correctly pruned tree of sandalwood [Source: Page et al (2012)]

Sandalwood grows with Mango, Jack fruit, Grape fruit, Rosewood, Coffee plant, Maccadamia plants as the long term host in Vietnam.

6.2.2 Weeding

Weeding is a very important management activity of the sandalwood plantation especially during the initial years of the plantation. If the weeds aren't controlled effectively during the initial years it may result in stunted growth of the plantation and even mortality and failure. The most common cause of plantation failure is inadequate weed control during the years of establishment. Newly planted seedling requires a weed-free area of at least 1 m^{2}for at least 3 years.

There are three forms of effective weed control (Page *et al* 2012)

• Manual pulling is used during the wet season, when conservation of soil moisture is not an issue.

• Mechanical cutting ('brushing') removal of woody weeds with a bush knife mainly during the dry season, when conservation of soil moisture is important.

• Chemical control with grass-selective or contact/ knockdown herbicides can be used, but these herbicides are generally too expensive and inaccessible for smallholders. Systemic broad spectrum herbicides (such as glyphosate) should be used with extreme caution. A systemic herbicide can move through the vascular system of the weeds and into the sandalwood through its haustoria, which can retard growth and potentially kill the sandalwood. Contact/knockdown herbicides may be used if spray drift is adequately controlled.

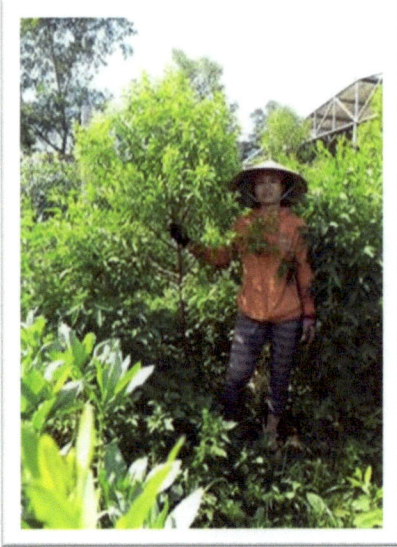

1year old plants in field

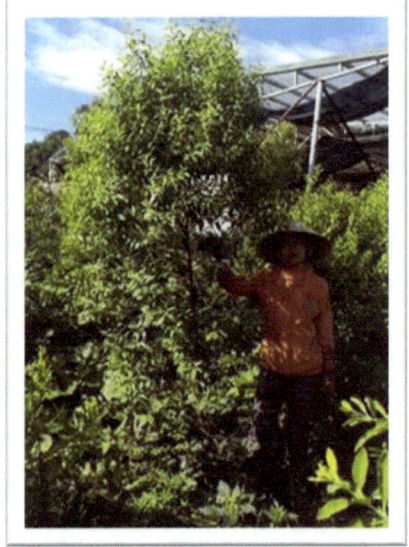

2 years old plants in field

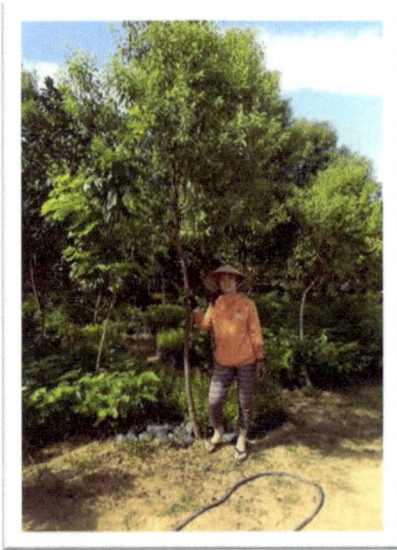

3 years old plants in field

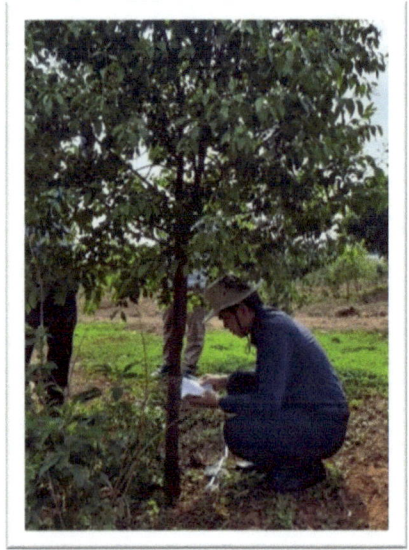

4 years old plants in field

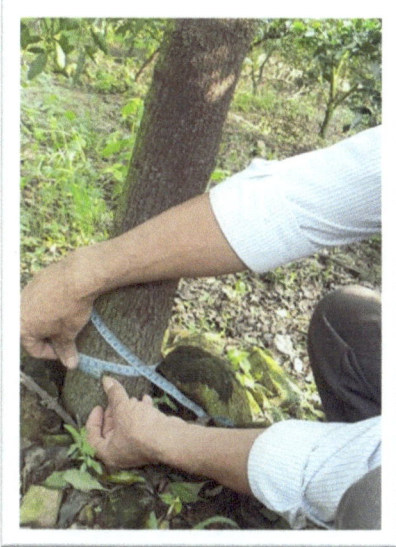

5 years old plants in field

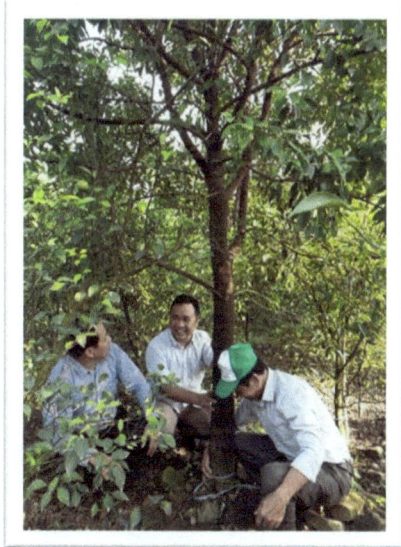

6 years old plants in field

Various plantations of Vietnam Sandalwood Group.

6.2.3 Management of Sandalwood to avoid damage to horticultural Hosts

Horticultural hosts play a very important role in sustainable management of the Sandalwood plantations as they provide intermediate income to the farmer. However, it has been observed in various plantations that after 5-6 years of growing along with the horticultural hosts, sandalwood tends to overpower the host reducing the yield drastically and in severe cases kill the host plants. It is suggested that to reduce the ill-effects of sandalwood on the horticultural host, following steps may be taken:

a) **Planting of permanent host:** A permanent host, say *Casuarina equisetifolia* may be planted about 1.5 feet away from sandalwood. This would provide easily available long term host for sandalwood till its rotation age and the roots of sandalwood would not have to reach the horticultural hosts for food and nutrition. *Casuarina equisetifolia* may be pruned after 1 year to a height of 4 feet and maintained as bush till the period of sandalwood rotation and harvested along with the main crop of sandalwood.

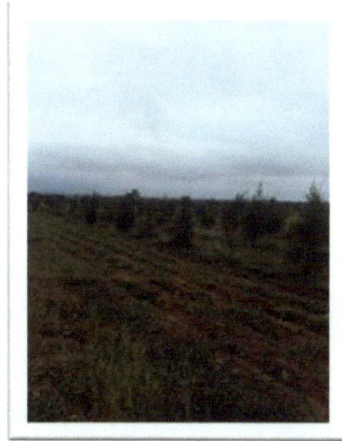

b) **Root pruning and Girdling of Sandalwood tree:** After attaining desired vegetative growth in sandalwood plants for 5-6 years, in order to reduce the power of sandalwood over the horticultural host and to induce stress in the plant, girdling and root pruning has to be carried out. Weakening of the sandalwood plants systematically would serve two purposes firstly, it would reduce the ill-effect of the sandalwood on horticultural host and secondly, it would induce stress in sandalwood plants initiating and hastening the process of heartwood formation. Root pruning and girdling has to be carried out as illustrated below.

GIRDLING   ROOT PRUNING

- **Girdling:** In order to induce stress in the sandalwood plants and to reduce the impact of sandalwood on the horticultural host, a small 1 inch strip of bark has to be removed from the main stem. The strip of the bark removed should be 50% of the girth of the tree and not the complete circumference. Remaining 50% of the stem can be girdled 6 months after the first one after has healed.

- **Root Pruning:** Dig a semi-circular 6-12 inches trench around the sandalwood plants to cut the active roots. The trenches should be dug only up to 50% of the circle around tree for first six months and second pruning in the remaining 50% of the circle in next 6 months.

CHAPTER 7

PESTS AND DISEASES OF SANDALWOOD IN NURSERIES AND PLANTATIONS AND THEIR MANAGEMENT

With an ever increasing demand of Sandalwood and its products in international markets large-scale plantation programs have necessitated a demand for planting stock, for which nurseries have been established across the world. However, systematic studies on Sandalwood nursery pests and diseases have been carried out in India. Some of the sandalwood pests encountered in India are responsible for heavy damage to the nursery stock. The same have been discussed in this chapter based on the published literature. Over 150 insects and pests have been reported in the sandalwood nurseries and plantations (Remadevi *et al* 2011). However, only few of them cause economic loss and they fall in the category of defoliators, sapsuckers, stem borers and termites. It is suggested that, a systematic study has to be carried out at local level to identify and control the pest problems encountered.

7.1 Insect pests of nurseries

i) Defoliators

The weevil *Sympiezomias spp* belonging to a polyphagous group was recorded as defoliating several trees of forestry importance including sandalwood seedlings in nurseries (Sivaramakrishnan et al. 1987). The weevil *Sympiezomias cretaceus* Faust (Curculionidae: Coleoptera), a polyphagous weevil feeds on the leaves from the edges towards the midrib. Sometimes the distal half of leaves gets cut off as a result of circular holes made in a line by the adult weevils. Feeding occurs at night. During the day adult beetles hide on the under surface of leaves or inside curled leaves or between webbed leaves.

Cryptothelea cramerii Westwood (Psychidae: Lepidoptera) cuts off the young seedlings, almost at ground level. A bag is constructed of small pieces, of the seedling stem, and these are placed side by side so as to form a cylindrical bag open at both ends. The sandalwood seedlings ultimately dry up under the attack of this pest.

The nymphs and adults of the grasshopper, *Holochlora albida* Kirby (Locustidae: Orthoptera) are green and are camouflaged as a leaf. They are well concealed among the young seedlings and they can only be detected when they move. These hoppers usually gnaw on tender shoots of sandalwood seedlings damaging the growing shoot and tender leaves of the plants.

Table 1. Name, family and order of defoliators present in nurseries and plant part affected by these insects. (Remadevi *et al* 2011)

Sl. No.	Name	Family	Order	Plant part affected
1	*Sympiezomias cretaceus*	Curculionidae	Coleoptera	Leaves
2	*Holochlora albida*	Locustidae	Orthoptera	Shoots
3	*Teratodes monticollis*	Acrididae	Orthoptera	Leaves
4	*Letana inflata*	Tettigonidae	Orthoptera	Leaves
5	*Cryptothelea cramerii*	Psychidae	Lepidoptera	Shoots & leaves
6	*Acanthopsyche moorei*	Psychidae	Lepidoptera	Leaves
7	*Pteroma plagiophleps*	Psychidae	Lepidoptera	Leaves

The bagworm, *Acanthopsyche moorei* Heyl (Psychidae: Lepidoptera) carries its bag upright at right angles to the stem or leaf of the seedling, but in the later instars the bag is heavy and is carried in a pendant position. The black caterpillar was seen defoliating the sandalwood seedlings giving a burnt appearance to the leaves. The attack of bagworm, *Pteroma plagiophleps* Hampson (Psychidae: Lepidoptera) results in total defoliation and drying up of sandalwood seedlings. Defoliation by these bagworms also imparts a burnt appearance to leaves.

ii) Sap suckers

The role of sap sucking pests belonging to the family Coccidae is very deleterious to the normal health, growth and reproduction of sandalwood plants.

The coccid, *Pulvinaria psidii* Mask is a common and destructive polyphagous insect in many sandalwood nurseries. The coccid, *P. maxima* Green is a new record on Sandalwood seedlings and is found very common on Neem growing in Arid zones. In the study carried out by Remadevi *et al* (2011) it was found to cause considerable damage to young sandalwood trees. The leathery pale brown adult females cover the tender shoots and stems in numbers and the white male scales are generally found conspicuously on the sandalwood leaves. The ovisacs are prominent and comparatively very long. The infestations lead to premature fall off of the leaves.

The coccid, *Ceroplastes actiniformis* Green has a thick pale white or pink waxy spherical test and the marginal area is divided into eight portions, which enclose a central cone area. Being sapsuckers, they cause severe sap drainage and sooty mold formation on the leaves below. The sap drainage leads to die back and ultimate death of sandalwood seedlings in nurseries. Spraying with monocrotophos (0.02–0.05%) kills this coccid.

The lac insect, *Tachardina lacca* Kerr. was observed on nursery plants and seedlings along with severe attack on trees. Quinalphos (Ekalux 20 AF) 0.5% or Dimethoate 0.2% mixed with 0.05% sticker was sprayed for effecting the control.

Feeding of sap by the coccid, *Inglisia bivalvata* Green caused browning of the leaves and withering. When the attack was severe, saplings succumb to the infestation. These scale insects are attached to twigs. Female scales look like bivalved shells. While adult males are winged, females are wingless and sedentary. Nymphs move out from underneath the scales of females and settle on tender branches to suck the sap. This insect occurs throughout the year. Parasites and predators help in the biological control of this pest. Spraying of 0.2–0.3% Chlorpyriphos or Quinalphos was effective in controlling this pest.

Pulvinaria psidii

P. *maxima*

Ceroplastes actiniformis

Inglisia bivalvata

Tachardina lacca

97

Table 2. Name, family and order of sapsuckers present in nurseries and plant part affected by these insects. (Remadevi *et al* 2011)

Sl No.	Name	Family	Order	Plant part affected
1	*Saissetia nigra*	Coccidae	Hemiptera	Shoots and leaves
2	*Saissetia coffeae*	Coccidae	Hemiptera	Shoots and leaves
3	*Pulvinaria psidii*	Coccidae	Hemiptera	Shoots and leaves
4	*Pulvinaria maxima*	Coccidae	Hemiptera	Shoots and leaves
5	*Ceroplastes actiniformis*	Coccidae	Hemiptera	Shoots and leaves
6	*Inglisia bivalvata*	Coccidae	Hemiptera	Shoots
7	*Tachardina lacca*	Coccidae	Hemiptera	Shoots
8	*Aspidiotus sp.*	Coccidae	Hemiptera	Leaves

Control measures of sap suckers.

Remadevi *et al* (2011) listed major pests of sandalwood and its control based on their experiments.

Table 3. Control of different sapsuckers on sandal.

Insect pests	Prevention/control measures
Ceroplastes ceriferus	Spraying of Monocrotophos (0.02–0.05%)
Saissetia sp.	Spraying 0.5% Quinalphos
Inglisia bivalvata	Spraying 0.2–0.3% Chlorpyriphos or Quinalphos
Tachardina lacca	0.5% Quinalphos (Ekalux 20 AF) along with 0.05% sticker sprayed thoroughly on the affected parts. Initial stages can be controlled by spraying 0.1% Rogor or 0.04% Cypermethrin.

7.2 Diseases of nurseries

Damping off and seedling wilt were the most serious diseases recorded during the survey in different states in India (Remadevi *et al* 2011).

i) Damping off

In most of the cases of seedling damping off in sandalwood it was found that species *of Fusarium, Rhizoctonia, Phytophthora* and *Pythium* were most commonly in infected seedlings. Most dangerous among the above was *Fusarium oxysporum* Schlecht which is the most virulent fungus found. It resulted in death of infected sandal seedlings by causing pre-emergence blight and vascular wilt. In vascular wilt, nematodes attacked seedlings along with *Fusarium* causing serious problems to seedlings. This fungus was found to spread rapidly in the tissues and the seedlings either wilt completely or rot off at ground line. If soil moisture is favorable, new roots develop and the seedlings continue to live though with less vigor and poorer growth. If such damage is followed by drought or by excess soil moisture, both of which discourage formation of new roots, seedling mortality may eventually occur.

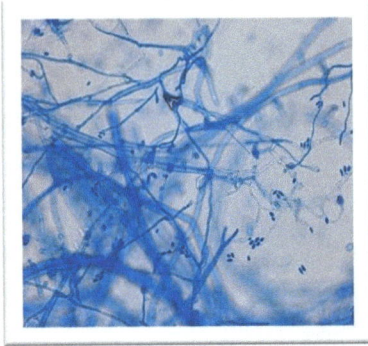

Spores of *Fusarium oxysporum* and damaged root in case of seedling damping off in Sandalwood

ii) Seedling wilt

Wilt is a systemic disease in sandal wood seedlings, where the entire individual or its parts exhibit wilting of the foliage in acropetal succession up to the shoot. The leaves become yellow, lose turgidity and fall off. The affected plant or the branch soon dies. Dwarfing, stunting and necrosis were also a common symptom found in seedlings. In all cases translocation of water and nutrients was adversely affected. Characteristic symptoms of vascular discoloration in the outer layers of the seedlings were evident. Main cause of seedling wilt is poor drainage of water coupled with over watering in the nursery medium which results in poor root growth and rotting of the roots.

Table 4. Control of different type fungal diseases (Remadevi *et al* 2011).

Type of disease	Casual organisms	Control measures
Fungus attack on seeds	*Fusarium, Alternaria, Aspergillus*	Store the de-pulped and dried sandal seeds with organomercuric compound (Cerasan/Agallol) dressing.
Seedling disease	*Fusarium or*	Selection of seeds, which are free of fungi, seed dressing with organomercuric compound.
1. Pre-emergence rot	*Phytophthora or*	Controlled watering, good drainage in containers.
2. Damping off	*Rhizopus* spp.	Drench the potting medium with copper fungicide and nematicide (Blitox/Bordeaux mixture and followed by nematodes
3. Fusarium wilt	*Fusarium or*	Quinalphos or Phorate). Controlled watering of plants
	Phytophthora species	

7.2 Insect Pests of plantations

a) Defoliators

Attack of defoliators in young plantations of Sandalwood are a serious concern for growers in Vietnam and worldwide. When defoliation is moderate and evenly distributed over the crown of a tree, the damage may not be easily observed. In many cases defoliation might reach 50% or more before the tree would appear abnormal. A single defoliation, though severe, rarely causes mortality. But in case of severe defoliation the defoliator attack reduces the immunity in plant and plant becomes susceptible to other pests. Defoliation reduces tree growth through the reduction of total photosynthetic area. With lower food-producing ability, the first parts of the tree to die are the small roots and twigs. If defoliation is heavy and repeated, death of the entire tree may follow. Occasionally outbreaks of defoliators arise rather suddenly and with no apparent warning. The major attacks occurs in May/June and a second, minor attack was observed in August/September. Initially, the only visible signs of damage were patches of defoliation at the tops of the affected trees and tiny pellets of excreta scattered beneath the tree.

Control measures of Defoliators

- Plough the field 2-3 times in December to bury the overwintering pupae in soil debris.

- Burn the fallen leaves which have been damaged by defoliators.

- The insecticides such as Ekalux 25 EC (quinalphos) @ 4 ml or Curacron 50EC (Profenofos) @ 2 ml in 1 litre of water should be sprayed in the affected fields in the evening to control the population of these defoliators. Best way of control of defoliators in precautionary sprays of insecticides and insect repellants in the month of February–March and August-September.

Defoliator attack in young plantation of Sandalwood in Ha Noi, Vietnam

b) Sap suckers and leaf miners

The majority of sap-sucking insects are in the orders Hemiptera (true bugs) and Homoptera (aphids, leaf and plant hoppers, and scales). Most of these insects are relatively small in size and injure the host in two ways:

1) Directly by sucking the host of part of its food supply and water, producing necrotic spots in host tissue, and

2) Indirectly by introducing plant diseases like sooty mould and fungal attacks.

The mouthparts of these insects and mites are formed into beak-like structures that are used to pierce host tissues and suck the sap. Damage by sap-sucking insects is often mistaken as a pathogen induced disease. A few of the sap-sucking insects are able to kill their hosts outright, but most reduce growth rates and weaken the tree. Trees injured by these insects may succumb to secondary insects or fungal diseases. Signs of sap-sucking insect injury consist of enlarged growths or galls, leaf curling, bleaching, or yellowing of foliage.

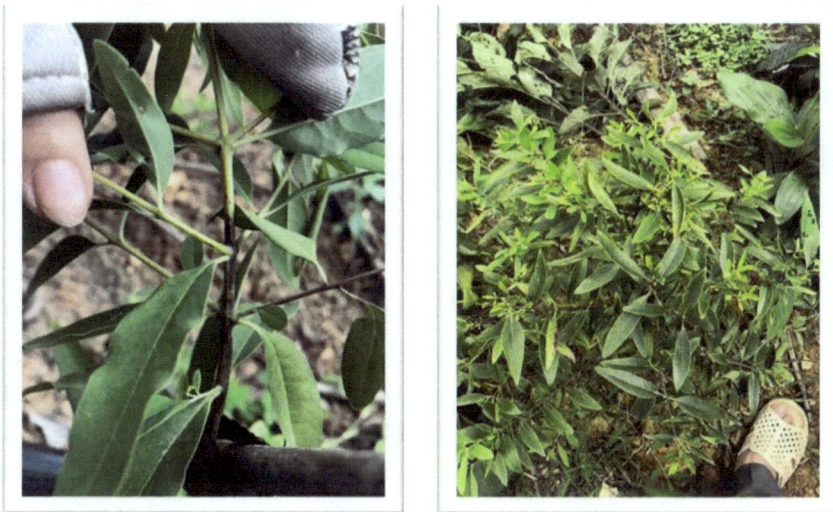

Sap sucker infestation in young plantation of sandalwood in Vietnam

Leaf miner on the other hand results in Leaf deformation - twisted or curled appearance. White or gray tunnels on leaf surface and stunted growth of the plant.

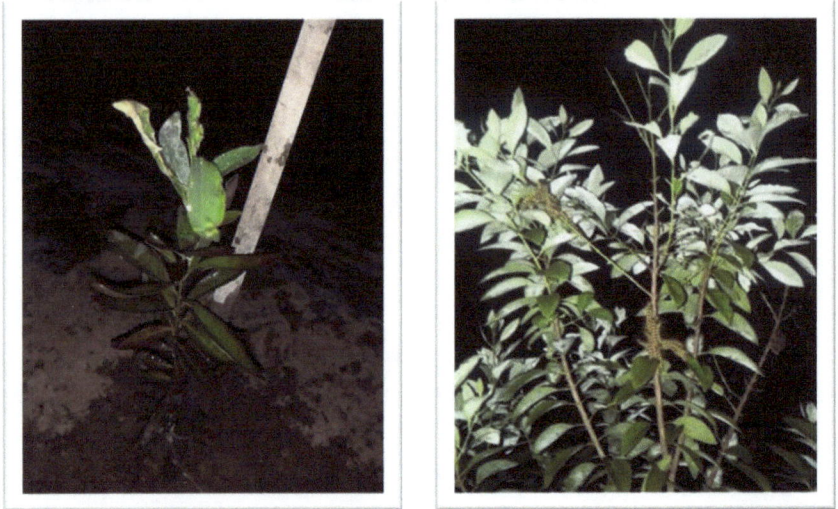

Leaf miner infestation in young plantation of sandalwood in Vietnam

Control measures

Spraying of Monocrotophos (0.02–0.05%), or Spraying 0.5% Quinalphos or Spraying 0.2–0.3% Chlorpyriphos or Quinalphos or 0.5% Quinalphos (Ekalux 20 AF) along with 0.05% sticker sprayed thoroughly on the affected parts. Initial stages can be controlled by spraying 0.1% Rogor or 0.04% Cypermethrin

c) Root nematode

Nematode attack facilitates the invasion by soil borne fungi and bacteria resulting in severe damage. Nematodes feed on below-ground parts of sandalwood and injure the root system. Recognizing nematode damage and symptoms can be problematic Notable aboveground symptoms of nematode damage are lack of vigor, twig dieback, and decline in growth and yield. Infestation to older trees may include chlorosis, and bark discoloration. Belowground symptoms include poor growth of feeder roots and main roots and soil adhering to roots. Root knot nematode infestations cause a characteristic swelling of the roots, called galls. The only way to confirm that observed symptoms are caused by nematodes is through close examination of the soil and/or root tissues.

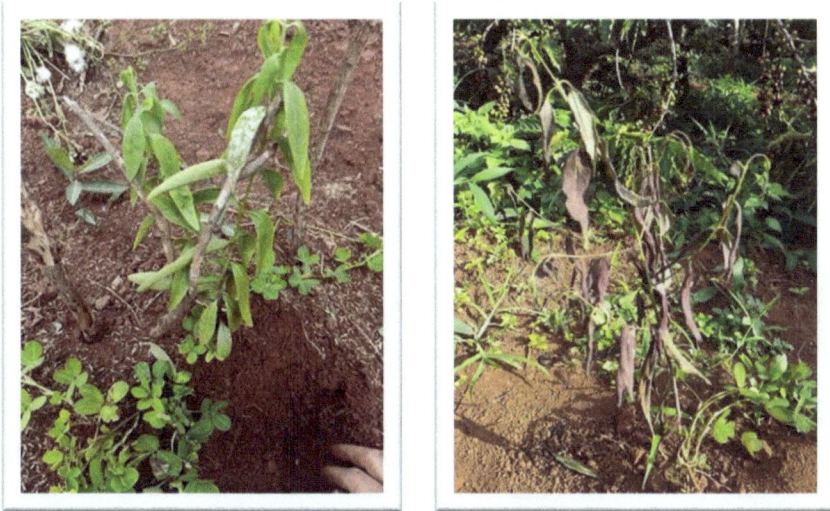

Root nematode infestation in young plantation of sandalwood in Vietnam

Control measures for nematodes

Management of nematodes is difficult. The most reliable practices are preventive, including sanitation and choice of plant varieties. You can reduce existing infestations through fallowing, crop rotation, and soil solarization. However, these methods reduce nematodes primarily in the top foot or so of the soil, so they are effective only for about a year. Try to provide optimal conditions for plant growth including sufficient irrigation and soil amendments to make plants more tolerant to nematode infestation.

Sanitation

Nematodes usually are introduced into new areas with infested soil or plants. Prevent nematodes from entering your garden by using only nematode-free plants purchased from reliable nurseries. To prevent the spread of nematodes, avoid moving plants and soil from infested parts of the garden. Don't allow irrigation water from around infested plants to run off, as this also spreads nematodes. Nematodes can be present in soil attached to tools and equipment used elsewhere, so clean tools thoroughly before using them in your garden.

Soil Solarization

You can use solarization to temporarily reduce nematode populations in the top 12 inches of soil, which allows the production of shallow-rooted annual crops and helps young woody plants become established before nematode populations increase. However, solarization won't provide long-term protection for fruit trees, vines, and woody ornamental plants.

Soil Amendments and Irrigation

You can add various organic amendments to the soil to reduce the effect of nematodes on crop plants. The amendments—which include peat, manure, and composts—are useful for increasing the water- and nutrient-holding capacity of the soil, especially sandy soils. Because nematodes more readily damage plants that are water-stressed, increasing the soil's capacity to hold water can lessen the effects of nematode injury. Likewise, more frequent irrigation can help reduce nematode damage. In either case, you will have just as many nematodes in the soil, but they will cause less damage.

Pesticides

Currently no chemical nematicides or soil fumigants are available to home gardeners for nematode control in plantations.

d) Stem borer

Stem borer operates at various stages of its life cycle damaging the trees in the plantation. Egg The eggs are dirty creamish white, oval, 7-8mm long and 3-3.2 mm wide. The grubs are creamy white in colour with elongated head of chestnut brown colour. The feeding of grubs starts during July and goes upto October. From October to February the grubs feed less and active feeding again starts in March. Adult Beetle 35 to 50 mm long, ashy grey with numerous black tubercles at the base of elytra and antennae are longer than body. In the older trees, grubs make 8-10 circuitous holes (for throwing out excreta and chewed wooden fibre) at an interval of 10-12 cm in the branches of main stem and reach to the trunk. The branches, stem and main trunk become hollow from inside. Ultimately the trees become weak and break in strong winds.

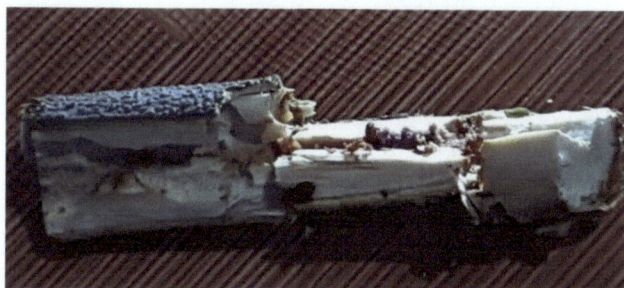

Stem borer infestation in young plantation of sandalwood in Dak Lak, Vietnam

Control measures for the shoot borer

1. Prune the infested branches in August-September before the entry of grub in the main stem.

2. Figure out all ejection holes of live infestation in each tree and plug them with wet clay, leaving only the lowermost hole untouched. Inject 2 ml of Dursban 20 EC (chlorpyriphos) saturated in kerosene oil in the lowest hole.

3. Avoid planting of other collateral hosts for shoot borers like mulberry, coffee in the vicinity of sandalwood plantations

e) Termites:

Termites are among the most successful groups of insects on Earth, colonizing most landmasses except Antarctica. Their colonies range in size from a few hundred individuals to enormous societies with several million individuals. Termite damage in forest nurseries and plantations is of widespread nature though the incidence and attack varies with species and locality. Subterranean damage to the root system of seedlings, tap roots and lateral roots of trees is very common. In nurseries it assumes serious proportions and often results in total loss. In nurseries, the affected plants show the signs of yellowing and wilting of apical leaves and ultimately death.

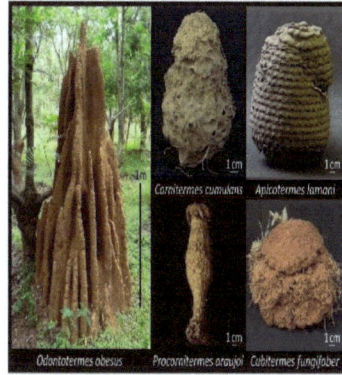

Termite infestation in young plantation of sandalwood in Ha Noi, Vietnam

Control of termites

- Drench the soil (around plants and before filling it in polythene bags) with Dursban 20 EC (Chlorpyriphos) @ 5 ml / litre of water.

- Destroy all the termitaria in the nursery or nearby areas.

- Flood the termitaria with Dursban 20 EC (Chlorpyriphos) using same dose as above.

7.4 Diseases of plantations.

Spectrum of diseases varies considerably in different agroclimatic regions in a country. Diseases primarily infest the nurseries and subsequently the plantations. In the nurseries diseases caused by soil borne fungi such as *Pythium, Phytophthora, Fusarium, Botryodiplodia* etc. cause rotting, damping off and collar rot. Several fungi like *Alternaria* spp., *Myrothesium roridum, Cercospora populina* etc. later on infects the leaves of plants in nurseries and plantations, causing leaf spots. Leaf blight, stem rot, stem canker, stem blight are more prevalent in the plantations.

a) Die back and Wilt

Symptoms of decline and dieback are often subtle, slow in developing, and usually uniform throughout the crown. A tree or shrub in the dieback stage, however, may have localized symptoms such as apparently healthy twigs and branches adjacent to dead or dying twigs and branches. Dieback usually begins in the top of a plant and progresses downward, but it may start on the lower branches, General symptoms of decline and dieback may include pale green or yellow leaves, delayed spring flush of growth, scorching of the leaf margins, small leaves, reduced twig and stem growth, early leaf drop, premature fall coloration, and, as the disease complex worsens, thinning of foliage in the crown, dieback of twigs and branches, and production of suckers on the branches and trunk

Wilt usually attacks older trees. The disease is characterized by yellowing of leaves from base of the trees upwards. Later on the leaves dry up and start to fall off leaving bare branches. The outer sap wood exhibits pink to reddish pink stain. Death of the tree occurs in a short span of few months.

Severely affected young plants of sandalwood by die back and wilt in Vietnam.

Control of Die back and wilt

Once the symptoms of decline or dieback are evident, it is difficult to stop or reverse the progress of disease. The key to control is early detection and prevention. The following measures will aid in maintaining the health of trees and shrubs.

- Maintain plant vigor by application of irrigation water in dry season and proper fertilization.

- Avoid sandalwood plantation in clayey soils with inadequate drainage.

- Avoid water logging.

- To avoid the spread of the pathogen through root contact, 3 feet deep trench should be dug around the infected tree.

- The wilted tree should be removed along with the roots and the soil around it should be fumigated with formalin (commercial grade) as soil drench (200ml commercial grade in 4 litres of water per square meter of land).

- Symptoms in wilting trees in initial stages (less than 10% chlorotic leaves) can be checked by applying systemic fungicides such as Bavistin @ 0.2% as soil drench.

b) Root rot

Signs of root rot in garden plants include stunting, wilting and discolored leaves. Foliage and shoots die back and the entire plant soon dies. If you pull up a plant with root rot, you will see that the roots are brown and soft instead of firm and white. Root rot is a disease that attacks the roots of trees growing in wet or damp soil. This decaying disease can cut the life short of just about any type of tree or plant and has symptoms similar to other diseases and pest problems, like poor growth, wilted leaves, early leaf drop, branch dieback, and eventual death. There are two causes for root rot, but the main cause is poorly drained or overwatered soils. These soggy conditions prevent roots from absorbing all the oxygen they require to live. As the oxygen-starved roots die and decay, their rot can spread to healthier roots, even if the soggy conditions have been rectified. Weakened roots are more susceptible to soil fungus, which is another cause of root rot. The fungus may be present but dormant in the soil for a long time; when the soil becomes waterlogged, the spores can come to life and attack the roots, causing them to rot and die. Some of the more well-known species of fungi that thrive in moist conditions and cause root rot are Pythium, Phytophthora, Rhizoctonia, and Fusarium.

Symptoms of root rot

Many symptoms of root rot mirror the signs of a pest infestation, which makes properly diagnosing it more difficult. The symptoms of root rot are obviously easier to spot above ground.

- Gradual or quick decline without an obvious reason.

- Stunted or poor growth.

- Small, pale leaves.

- Wilted, yellowed, or browned leaves.

- Branch dieback.

- Thinning of the canopy.

- On some species, the fungus grows up from the roots in the inner bark and causes cankers, or sunken dead areas.

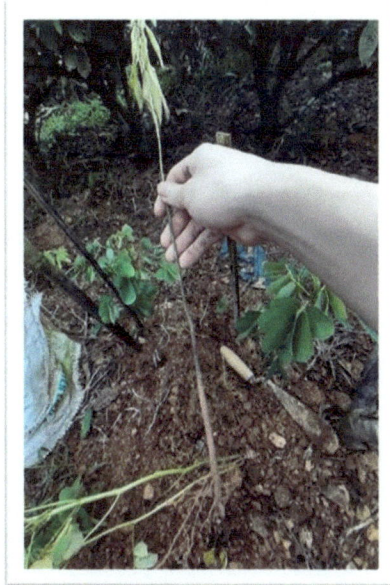

Root rot infestation in young plantation of sandalwood in Dien Bien Vietnam

Control of root rot

- Tree root diseases are best controlled by prevention. If you are considering planting new trees, choose disease-resistant varieties or cultivars, only plant in well-drained soil, and avoid overwatering. You can also create irrigation moats to keep the water from pooling against tree trunks.

- Moderately affected trees can sometimes be saved early on by pruning out the infected roots. Always be sure to disinfect any tools you work with before using them again. If a tree is significantly infected, the best way to control it from spreading the disease to healthier trees is to remove it entirely.

- Chemicals, such as chloropicrin or methyl bromide, won't completely cure the disease but can reduce the level of the infection. These fumigants are applied in and around the base of the infected trees or in holes left after trees have been removed.

c) Leaf spots

The spots start from margin in most of the cases as light brown coloured lesion with broad dark chocolate brown border. Spots contain very minute dark green coloured sporodochia, which appear as raised dots. The attack is mostly on the lower side of leaf, where greyish-green discolouration is visible. The lesion under humid conditions are covered with greenish growth of saprophytic fungus. The affected portion of the leaf turns necrotic and dies giving a blight end appearance. The disease symptoms were characterized by spots, which were pale brown to grey with dark margins, and the pale portion had black coloured pycnidial bodies. Spots resulted in heavy premature defoliation.

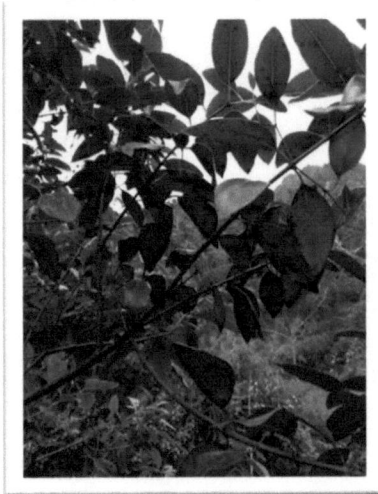

Leaf spot infestation in young plantation of sandalwood in Dien Bien Vietnam

Control measures of leaf spot

2-3 sprays should be given with Bavistin 50 WP (1 g per litre of water) followed by Tilt 25 EC (1 ml per litre of water) at fortnightly interval starting from last week of June to avoid leaf spots. Drenching with Bavistin 50 WP @ 0.2 per cent should be done using 25 litre of water per tree during second fortnight of August.

d) Powdery mildew

Appearing as white or grayish powder-like patches on the surface of leaves and stems, powdery mildew develops mostly on newer growths. Although the fungal spores often look like someone sprinkled baby powder on foliage, the disease can also appear as blotchy, felt-like mats or cobweb-like formations.

Disease symptoms increase when cooler weather is paired with high humidity, and these signs occur more frequently on areas of a tree that are shaded and lack good air flow. As this mildew spreads, leaves begin to yellow and wilt, and eventually, the entire branch dies. Advanced symptoms of an infection also include distorted leaves, premature leaf drop, and blemishes on fruit.

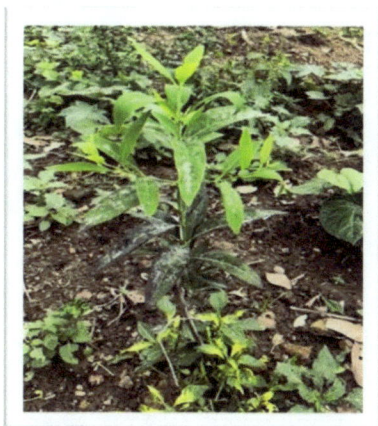

Powdery mildew infestation in young plantation of sandalwood in Ha Noi Vietnam

Control measures of Powdery mildew

- To prevent conditions in which powdery mildew can thrive, don't plant trees in extremely shaded areas. Opt for locations with good air flow, sunlight, and plenty of growing room. If necessary, prune trees to provide better air circulation.

- Proper sanitation is important to reduce the spread of infection. Immediately remove and dispose of infected leaves or branches that have dropped. Don't use them for composting or mulching. And be sure to clean and disinfect tools after pruning before you use them again.

- Fungicides can be used to control infestations, though many cases of powdery mildew do not require this drastic measure. Because the mildew spreads rapidly, fungicides must be applied at the first sign of the disease's symptoms. Be sure to prune out severely infected branches first. Sulfur can be used as a spray or dust, but can cause other kinds of damage to the foliage—especially in high temperatures–if not applied carefully.

7.4 Nutrient deficiencies of plantations

Nutrient deficiency in plants is responsible for below average growth in the plants which can seriously hamper the final yield expected out of a crop cycle. Discoloration of leaves is an effective way to assess the nutrient deficient in the growing plants. A guide of nutrient deficiency symptoms is presented below for general identification in sandalwood plantations. It may also be mentioned here that sometimes the soil conditions also affect the nutrient availability (Soil pH, drainage status) so one needs to first ascertain that all other soil conditions are favourable for nutrient uptake before using the leaf appearance as tool to identify the nutrient deficiency.

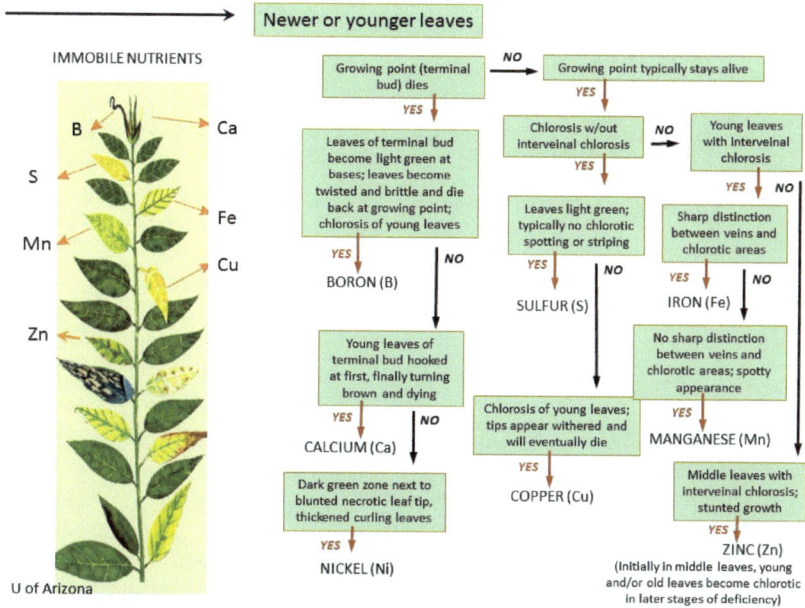

Newer or younger leaves

IMMOBILE NUTRIENTS

Growing point (terminal bud) dies — NO → Growing point typically stays alive

YES ↓

Leaves of terminal bud become light green at bases; leaves become twisted and brittle and die back at growing point; chlorosis of young leaves

YES ↓ / NO →

BORON (B)

Young leaves of terminal bud hooked at first, finally turning brown and dying

YES ↓ / NO →

CALCIUM (Ca)

Dark green zone next to blunted necrotic leaf tip, thickened curling leaves

YES ↓

NICKEL (Ni)

Chlorosis w/out interveinal chlorosis — NO → Young leaves with interveinal chlorosis

YES ↓

Leaves light green; typically no chlorotic spotting or striping

YES ↓ / NO →

SULFUR (S)

Chlorosis of young leaves; tips appear withered and will eventually die

YES ↓

COPPER (Cu)

Sharp distinction between veins and chlorotic areas

YES ↓ / NO →

IRON (Fe)

No sharp distinction between veins and chlorotic areas; spotty appearance

YES ↓

MANGANESE (Mn)

Middle leaves with interveinal chlorosis; stunted growth

YES ↓

ZINC (Zn)
(initially in middle leaves, young and/or old leaves become chlorotic in later stages of deficiency)

U of Arizona

Source: http://landresources.montana.edu/soilfertility/nutrientdeficiencies.html

CHAPTER 8

HEARTWOOD FORMATION AND HARVESTING IN SANDALWOOD

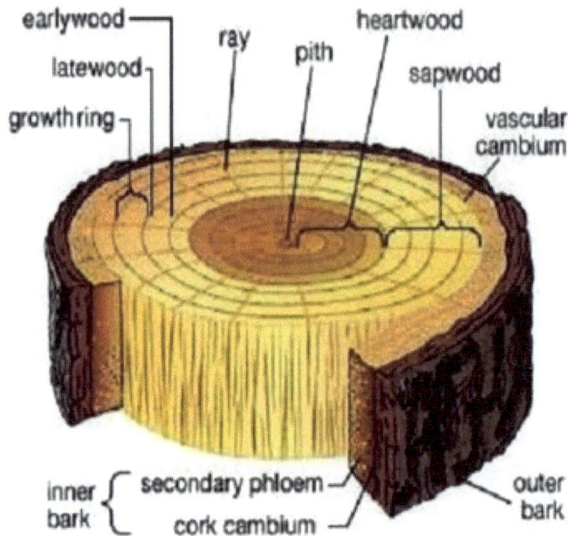

8.1 Heartwood formation in vascular plants

Wood formation is a complex biological process, involving five major developmental steps, including:

a) Cell division from a secondary meristem called the vascular cambium.

b) Cell expansion (cell elongation and radial enlargement).

c) Secondary cell wall deposition,

d) Programmed cell death, and

e) Heartwood formation.

Heartwood formation is a process that involves the death of older cells in the tree and it gets deposited at the central portion of the trunk. It is well established fact that heartwood formation is a stress related phenomenon where in stress in the plants either natural or induced results in its formation.

Heartwood formation in sandal trees generally starts around 7 years of age. Certain factors, generally relating to stress, such as gravelly dry soil, insolation, and range of elevation (500-700 m), seem to provide the right environment for the formation of heartwood, irrespective of the size of the stem after 10 years of age. The occurrence of heartwood varies. The value of heartwood is due to its oil content, and the superiority of the oil is due to the percentage of santalol. In a tree the oil content is highest in the root, next highest in the stem at ground level, and gradually tapers off towards the tip of the stem. Similarly, there is a gradient in oil content from the core to the periphery of the heartwood in a stem (Page *et al* 2012).

8.2 Sandalwood harvesting

The time needed to develop enough heartwood for harvest will vary between trees and growing environments. Tree size is a good indication of when the tree is ready for harvest. The minimum size at which a tree can be harvested is a trunk diameter of 15 cm at breast height, which corresponds to a tree with a basal diameter of about 20 cm. Under good growing conditions, a tree of this size is approximately 15–20 years old. Oil is particularly concentrated in the roots and butt of the sandalwood tree; therefore, the stump and roots need to be dug out to gain maximum saleable value from the tree. Heartwood is present in the branches of older trees but is unlikely to be in the branches of planted sandalwood with a rotation of 15–20 years.

8.2.1 Preliminary processing

Sandalwood is sold by weight. Before selling it, the only requirement is to remove the sapwood from around the heartwood. This is usually done by gradually cutting the sapwood away with a bush knife. This process results in two waste products: pure sapwood chips and second cutting chips (2CC), which are both used in the manufacture of incense. The 2CC are chips that retain some heartwood.

8.2.2 Sandalwood processing for carving, agarbatti and oil

The carving log attracts the highest price in the marketplace, followed by oil then powder. All products require the outer sapwood to be removed (de-sapped), which is typically done by the harvester or grower before sale. The total cost of processing is lowest for carving logs; they require further de-sapping, which removes the final layer of sapwood next to the heartwood and produces second cutting chips (2CC). The log ends are then sealed (usually with wax) to prevent rapid drying and cracking. Powdered wood requires additional milling and blending of different powders to achieve a final product that is acceptable for use in agarbatti. Sandalwood oil is the most expensive to produce because the heartwood needs to be powdered before the oil is extracted by an energy intensive and complex process called distillation. During distillation, steam is generated in a boiler, which is typically fuelled by diesel, coconut oil, wood or electricity, and passes through powdered heartwood, where it mixes with the heartwood oil. This mixture is then cooled and condensed, with the oil forming a layer on top of the water. Distillation can take several days to liberate all the oil from the heartwood. The production of high-quality oil requires a high level of experience and knowledge of the process. Wood extracted from a 15 year old plantation on an average gives 3-4% oil w/w basis.

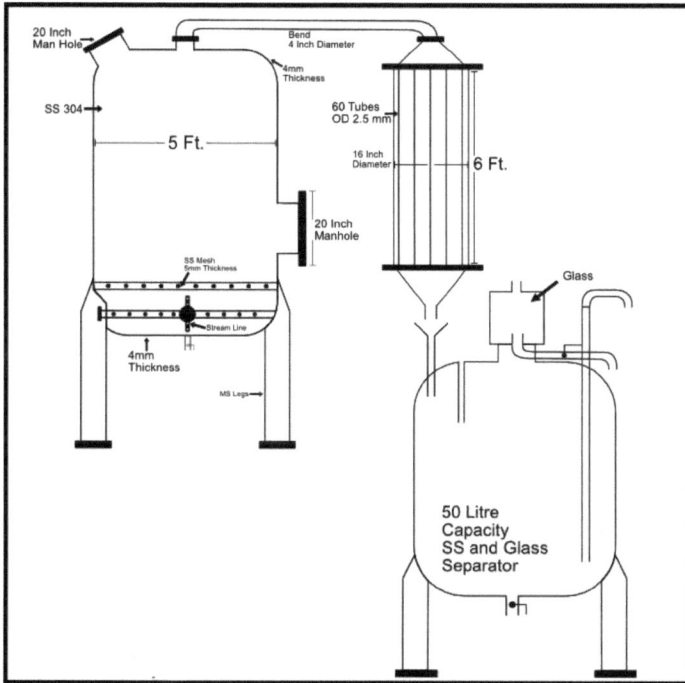

20 Inch
Man Hole

Bend
4 Inch Diameter

4mm
Thickness

SS 304

5 Ft.

60 Tubes
OD 2.5 mm

16 Inch
Diameter

6 Ft.

20 Inch
Manhole

SS Mesh
5mm Thickness

Stream Line

4mm
Thickness

MS Legs

Glass

50 Litre
Capacity
SS and Glass
Separator

REFERENCES

- Ananthapadmanabha H.S., Nagaveni H.C. & Rai S.N. (1988). Dormancy principles in sandal seeds (*Santalum album* Linn.). Myforest 24: 22 – 24.

- Ananthapadmanabha, H.S. 2000. Sandalwood and its marketing trend. My Forest 36: 147-151.

- Ananthapadmanabha H.S.and Gowda, V.V.S (2011). Sustainable Supply of Global Sandalwood for Industry In Proceedings of the Art and Joy of Wood conference, 19-22 October 2011, Bangalore, India 39-46p.

- Angadi, V.G., Kamal, B.S. and Rai, S.N. 1988. Effect of deficiency of trace elements on leaf area, chlorophyll level, and photosynthetic efficiency in tree seedlings. My Forest 24(2): 124-128.

- Annapurna D, Rathore, T.S. and Somashekhar, P.V. 2005. Impact of clones in a clonal seed orchard on the variation of seed traits, germination and seedling growth in *Santalum album* L. Silvae Genetica,; 5:153-160.

- Annapurna D, Rathore TS, Geeta Joshi (2006). Modern nursery practices in the production of quality seedlings of Indian sandalwood (*Santalum album* L.- stage of host requirement and screening of primary host species. Journal of Sustainable Forestry 22: 33- 44.

- Anon (1992) Sandal (*Santalum album* L). A monograph. (eds) Srinivasan , ICFRE publication

- Anon. 2005. Annual Report. Karnataka Forest Department, Bangalore.

- Anon (2018). Special Feature: Australia's Would-Be King of Sandalwood. An article published in Magazine Web Edition of Hindustan Today of April-June.

- Applegate, G.B., Chamberlina, J.C., Daruhi, G., Feigelson, J.L., Hamilton, L., Mc Kinnell, F.H., Neil, P.E., Rai, S.N., Rodehn, B., Statham, P.C. and Stemmermann, L. 1990. Sandalwood in the pacific: a state-of-knowledge synthesis and summary. Pp 1-11. In: Hamilton, L. and Conrad, C.E. (eds.) Proceeding of the symposium on Sandalwood in the Pacific, Honolulu, Hawaii, 9-11 April 1990. USDA Forest Service Gen Tech Rep PSW-122.

- Arun Kumar, A.N., Joshi, G. and Mohan Ram, H.Y. 2012. Sandalwood: history, uses, present status and the future Current Science 103(12): 1408-1416.

- Arun Kumar, A. N., Joshi, G., Rao, M. S., Rathore, T. S. and Ramakantha, V. (2016). The population decline of Indian sandalwood and people's role in conservation – an analysis. In: Nautiyal S., Schaldach R., Raju K.V, Kaechele H., Pritchard B., Rao K.S. (eds) Climate Change Challenge (3C) and Social-Economic, Ecological Interface-Building. Springer International Publishing, Switzerland, pp. 377–387.

- Arunkumar, A. N, Warrier, Rekha & Arunachalam, Shanthi & Joshi, Geeta (2018). Allozyme Variations to Measure Genetic Diversity in Clonal Accessions of Indian Sandalwood (*Santalum album*). 4. 10.20431/2454-9487.0401001.

- Azeez S A, Nelson R, Prasadbabu A and Rao MS (2009). Genetic diversity of *Santalum album* using random amplified polymorphic DNAs.Vol 8 No 13.

- Bahuguna, A. (2007). Conservation genetics: A management tool. Annals of Forestry, 15: 159-200.

- Barrett, D.R. 1988. *Santalum album* (Indian Sandalwood) literature survey. Mulga Research Center, Western Australia.

- Baskin C.C. & Baskin J.M. (1998). Seeds: Ecology, Biogeography, and Evolution of Dormancy and Germination. Academic Press, San Diego, USA.

- Beniwal B.S. & Singh N.B. (1989). Observations of the flowering, fruiting and germination behaviour of some useful forest plants of Arunachal Pradesh. Indian Forester 115: 216 – 227.

- Boardman R, Bush D, Butcher T , Harwood C, Spencer D and Stackpole D (2002) Australian Low Rainfall Tree Improvement Group: Compendium of Softwood Tree Improvement Strategies. A report for the RIRDC/ Land & Water Australia/ FWPRDC, Joint Venture Agroforestry Program Supported by the Natural Heritage Trust. 55p.

- Braun, N.A., Sim, S., Kohlenberg, B., Lawrence, B. M. (2014). Hawaiian Sandalwood: Oil composition of *Santalum paniculatum* and comparison with other sandal species. Natural Product Communications, 9(9), 1365–1368.

- Byrne, M., MacDonald, B., Broadhurst, L. and Brand, J. 2003. Regional genetic differentiation in Western Australian sandalwood (*Santalum spicatum*) as revealed by nuclear RFLP analysis. Theor Appl Genet 107: 1208–1214.

- Chauhan, S.S. and Aggarwal P. 2007. Growth models for *Santalum album* L. Pp. 153-157. In: Gairola, S., Rathore, T.S., Joshi, G., Arun Kumar, A.N. and Aggarwal, P.K. (eds.) Conservation, Improvement, Cultivation and Management of Sandal (*Santalum album* L.). IWST, Bangalore.

- Cheng, Qingwei & Yueya, Zhang & Niu, Meiyun & Yan, Haifeng & Xinhua, Zhang & Teixeira da Silva, Jaime & Guohua, Ma. (2017). Limitations in the tissue culture of Indian sandalwood tree (*Santalum album* L.). Advances in Biotechnology. 13 pp.

- Clarke B. & Doran J. (2012). Review of *Santalum album* L. seed pre-germination treatment with a focus of low cost methods. Proceedings of the International Sandalwood Symposium, 21 – 24 October, 2012, Honolulu, Hawaii.

- Cromer E.L. & Woodall G. (2007). Breaking mechanical dormancy in quandong using silica gel and enhancing germination response using gibberellic acid. Alcoa World Alumina Australia 28: 1 – 10.

- De Groot, A. C., Schmidt, E. (2016). Essential oils, part VI: Sandalwood oil, ylang-ylang oil, and jasmine absolute. Dermatitis, 27(4), 170–175.

- Dhanya B, Viswanath S, Purushothaman S (2010) Sandal (*Santalum album* L.) conservation in southern India: A review of policies and their impacts. Journal of Tropical Agriculture 48 (1-2): 1-10

- Dileepa M.M. Jayawardena M. , Jayasuriya K.M.G. Gehan and Walck Jeffrey (2015). Confirmation of morphophysiological dormancy in sandalwood (*Santalum album*, Santalaceae) seeds. J.Natn.Sci.Foundation Sri Lanka 2015 43 (3): 209-215.

- Doran, J., Thomson, L., Brophy, J., Goldsack, B., Bulai, P., Faka'osi, T and Mokoia, T. 2005. Variation in heartwood oil composition of young sandalwood trees in the South Pacific (*Santalum yasi, S. album* and F1 hybrids in Fiji, and *S. yasi* in Tonga and Niue). Sandalwood Research Newsletter 20: 3-7.

- Doran, J. (2012), Review of *Santalum album* seed pre germination treatments with a focus on low cost methods. Proceedings of international sandalwood symposium, 21-24 october, Honolulu, Hawaii.

- Edward F. Gilman and Amanda Bisson (2019) Developing a Pruning Program for Young Trees. University of Florida IFAS Extension,. Publication No. ENH 1062

- Fischer, C.E.C. 1928. The original home of *Santalum album* L. Journal of Indian Botanical Society 7: 12-13.

- Fischer, C.E.C. 1938. Where did the sandalwood tree (*Santalum album* L.) evolve? Journal of Bombay Natural History Society 40: 458-466.Fox, J.E. 2000. Sandalwood: the royal tree. Biologist (London) 47: 31-34.

- Gairola, S., Ravi Kumar, G. and Aggarwal, P.K. 2007. Status of production and marketing of Sandalwood (*Santalum album* L.). Pp. 1-8. In: Gairola, S., Rathore, T.S., Joshi, G., Arun Kumar, A.N. and Aggarwal, P.K. (eds.) Conservation, Improvement, Cultivation and Management of Sandal (*Santalum album* L.). IWST, Bangalore.

- Gamage Y.M.M., Subasinghe S.M.C.U.P. & Hettiarachchi D.S. (2010). Change of seed germination rate with storage time of *Santalum album* L. (Indian sandalwood) seeds. Proceedings of the 15th International Forestry and Environment Symposium (eds. D.M.H.S.K. Ranasinghe, B.M.P. Singhakumara, H.S. Amarasekera, N.J.G.J. Bandara, U.A.D.P. Gunawardana & S.M.C.U.P. Subasinghe), volume 15, Department of Forestry and Environmental Science, University of Sri Jayewardenepura, pp. 279 – 281.

- Ghosh, S.K., Balasundaran, M. and Ali, M.I.M. 1985. Studies on spike disease of Sandal. KFRI Research Report No. 37. Kerala Forest Research Institute, Peechi, Thrissur. 56 p.

- Ghosh Dasgupta, Modhumita & Ulaganathan, Kandasamy & Dev, Suma & Balakrishnan, Swathi. (2019). Draft genome of *Santalum album* L. provides genomic resources for accelerated trait improvement. Tree Genetics & Genomes. 15. 10.1007/s11295-019-1334-9.

- Gowda V.V.S (2011) Global Emerging Trends on Sustainable Production of Natural Sandalwood In Proceedings of the Art and Joy of Wood conference, 19-22 October 2011, Bangalore, India. 3-12pp.

- Hirano R.T. (1990). Propagation of *Santalum*, sandalwood tree. USDA Forest Service General Technical Reports 122: 43 – 45.

- Indriokoa Sapto and RatnaningrumYeni W N (2015) Habitat loss caused clonality, genetic diversity reduction and reproductive failure in *Santalum album* (Santalaceae), an endangered endemic species of Indonesia. Procedia Environmental Sciences 28 (2015) 657 – 664.

- Iyengar, A.V.V. 1960. The relation of soil nutrients to the incident of spike disease in sandalwood (*Santalum album* Linn). Indian Forester 86(4): 220-230.

- Jain, S.H., Angadi, V.G., Ravikumar, G., Theagarajan, K.S. and Shankaranarayana, K.H. 1999. Strudies on cultivation and chemical utilization of sandal (*Santalum album* L.). Fafai Journal 3: 49-53.

- Jain, S.H., Angadi, V.G., Shankaranarayana, K.H. and Ravikumar, G. 2003. Relationship between girth and percentage of oil in sandal provenances. Sandalwood Research Newsletter 18: 4-5.

- Jain, S.H., Angadi, V.G. and Shankaranarayana, K.H. 2007a. Edaphic, environmental and genetic factors associated with growth and adaptability of sandal (*Santalum album* L.) in provenances. Sandalwood research newsletter 17: 6-7.

- Jain, S.H., Arya, R. and Kumar, H. 2007b. Distribution of sandal (*Santalum album* L.) current growth rates, predicted yield of heartwood and oil content and future potential in semi arid and arid region of Rajasthan, India. Forests, Trees and Livelihoods 17: 261-266.

- Jeeva, V., Saravanan, S., Devaraj, P. and Lakshmidevi, R. 1998. Malady and remedy of sandal cultivation in farmlands and private lands-An overview. In: Radomiljac, A.M., Ananthapadmanabha, H.S., Welbourn, R.M. and Rao, K.S. (eds.) Sandal and its products. ACIAR Proceedings No. 84. ACIAR, Canberra, Australia.

- Joshi, G. and Arun Kumar. A.N. 2007. Standardization of optimum conditions for storage of *Santalum album* L. seeds for ex situ germplasm conservation. Pp. 52-54. In: Gairola, S., Rathore, T.S., Joshi, G., Arun Kumar, A.N. and Aggarwal, P.K. (eds.) Conservation, Improvement, Cultivation and Management of Sandal (*Santalum album* L.). IWST, Bangalore.

- Kemp, R. H., Namkoong, G. and Wadsworth, F. H. (1993). The nature of forest genetic resources. In: Conservation of genetic resources in tropical forest management principles and concepts. FAO Forestry paper 107. Rome, pp. 514.

- Krishna Bahadur KC. (2019)Status and distribution of Sandalwood (*Santalum album*) in Nepal: A study of Pyuthan district. Species, , 20, 13-23

- Kumaravelu, G., Bharathi, R.K., Jainludeen, A. and Balraj, V.C. 2007. Resurrection of the rare aromatic *Santalum album* population in Tamil Nadu. Pp. 14-22. In: Gairola, S., Rathore, T.S., Joshi, G., Arun Kumar, A.N. and Aggarwal, P.K. (eds.) Conservation, Improvement, Cultivation and Management of Sandal (*Santalum album* L.). IWST, Bangalore.

- Loveys B.R. & Jusaitis M. (1994). Stimulation of germination of quandong (*Santalum macuminatum*) and other Australian native plant seeds. Australian Journal of Botany 42: 565 – 57.

- Majupuria, T.C. and Joshi, J.P. 1988. Religious and useful plants of India and Nepal. Craftsman Press, Bangkok.

- Martin A.C. (1946). The comparative internal morphology of seeds. American Midland Naturalist 36: 513 – 660. DOI: http://dx.doi.org/10.2307/2421457.

- Mathur, C.M. 1961. Artificial regeneration of *Santalum album* in Rajasthan. Indian Forester 87(1): 37-39.Mc Kinnell, F.H. 1990. Status of management and silviculture research on sandalwood in eastern Australia and Indonesia. Pp. 19-25. In: Hamilton, L. and Conrad, C.E. (eds.) Proceeding of the Symposium in the Pacific. Honolulu, Hawaii, 9-11 April 1990. USDA Forest Service Tech. Rep. PSW-122.

- Nagaveni H.C. & Srimathi R.A. (1980). Studies on germination of the sandal seeds, *Santalum album* Linn. II. chemical stimulant for germination. Indian Forester 106: 792 – 799.

- Nageswara Rao, M. 2004. Mapping genetic diversity of sandal (*Santalum album* L.) genetic resources in peninsular India using biochemical and molecular markers: lessons for *in-situ* conservation. Ph.D. thesis, Forest Research Institute University, ICFRE, Dehra Dun, India.

- Nageswara Rao, M., Ganeshaiah, K.N. and Uma Shaankar, R. 2007. Assessing threats and mapping sandal (*Santalum album* L.) resources in peninsular India: identification of genetic hot-spot for *in-situ* conservation. Conservation Genetics 8: 925-935.

- Namkoong Gene, Hyun, C, Kang, Jean S (2012) Tree Breeding: Principles and Strategies: Springer Publication 174p.

- Neil, P.E. 1986. Sandalwood in Vanuatu. Forest Research Report 5/86, Vanuatu Forest Service, Vanuatu.

- Neil, P.E. 1990a. Growing sandal wood in Nepal- potential silviculture methods and research priorities. Pp. 72-75. In: Hamilton, L. and Conrad, C.E. (eds.) Proceeding of the Symposium on Sandalwood in the Pacific. Honolulu, Hawaii, 9-11 April 1990. USDA Forest Service Tech. Rep. PSW-122.

- Neil, P.E. 1990b. Possible techniques for raising and planting sandalwood in Nepal. Banko Janakari 2(3): 223-228.Radomiljac, A.M. and Mc Comb, J.A. 1998. Nitrogen-fixing and non-nitrogen fixing woody host influences on the growth of the root hemi-parasite *Santalum album* L. Pp. 54-57. In: Radomiljac, A.M, Ananthapadmanabha, H.S., Welbourn, R.M. and Rao, K.S. (eds.) Sandal and its products. ACIAR Proceedings No.84. ACIAR, Canberra, Australia.

- Nikam T.D. & Barmukh R.B. (2009). GA$_3$ enhances *in vitro* seed germination in *Santalum album*. Seed Science and Technology 37: 276 – 280.

- Nurchahyani, Yeni & Indrioko, Sapto & Faridah, Eny & Syahbudin, Atus. (2017). The effects of population size on genetic parameters and mating system of sandalwood in Gunung Sewu, Indonesia. Indonesian Journal of Biotechnology. 20. 182. 10.22146/ijbiotech.24347.

- Orwa C, Mutua A , Kindt R , Jamnadass R, Simons A. 2009. Agroforestree Database:a tree reference and selection guide version 4.0 (http://www.worldagroforestry.org/af/treedb/)

- Page, Tony & Tate, Hanington & J, Tungon & M, Tabi & P, Kamasteia. (2012). Vanuatu Sandalwood: Growers' guide for sandalwood production in Vanuatu. ACIAR Publication Government of Australia.56p.

- Paliwal R.L.,(1956) Morphological and embryological studies in some Santalaceae. Agra University Journal Research,; 5: 193-284.

- Pallavi, Aparna (2018), In Return of the scented wood, News paper article in Down to Earth daily from Bangalore Edition dated 19th September 2018.

- Prasetyaningtyas M. (2007). *Santalum album* L. Seed Leaflet No. 116, pp. 1 – 2. Forest and Landscape Denmark, Hørsholm, Kongevej, Denmark.

- Rai, S.N. and Sharma, C.R. 1986a. Relationship between height and diameter increment of sandal (*Santalum album*). Van Vigyan 24(3&4): 105-108.

- Rai, S.N. and Sharma, C.R. 1986b. Study of diameter growth in Sandal. Journal of Tropical Forestry 2: 201-206. Rai, S.N. 1990. Status and Cultivation of Sandalwood in India. Pp. 66-71. In: Hamilton, L. and Conrad, C.E. (eds.) Proceeding of the Symposium on Sandalwood in the Pacific. Honolulu, Hawaii, 9-11 April 1990. USDA Forest Service Tech. Rep. PSW-122.

- Rai SN (1990). Status and cultivation of sandalwood in India. In: Proceeding of the Symposium on Sandalwood in the Pacific. Hamilton L, CE Conrad (Ed.), USDA Forest Service General. Technical Paper PSW-122, Honolulu, USA pp 66-71.

- Rajan, N.M and Jayalakhsmi S (2017). Predictive Species Habitat Distribution Modelling of Indian Sandalwood Tree Using GIS. Pol. J. Environ. Stud. Vol. 26, No. 4 (2017), 1627-1642.

- Rangaswamy N.S. & Rao P.S. (1963). Experimental studies on *Santalum album* L. -establishment of tissue culture of endosperm. Phytomorphology 13: 450 – 454.

- Rao, N.M., Ravikanth, G., Ganeshaiah, K.N., Rathore, T. S. and Uma Shaanker, R. 2007. Assessing threats and identifying the ecological niche of sandal resources to identify 'hotspots' for in situ conservation in Southern India. Pp. 23-31. In: Gairola, S., Rathore, T.S., Joshi, G., Arun Kumar, A.N. and Aggarwal, P.K. (eds.) Proceedings of the National Seminar on Conservation, Improvement, Cultivation and Management of Sandal (*Santalum album* L.). IWST, Bangalore.

- Ratha Krishnan, P., Rajwant K. Kalia, Tewari, J.C. and Roy, M.M. (2014). Plant Nursery Management: Principles and Practices. Central Arid Zone Research Institute, Jodhpur, 40 p

- Remadevi O.K, Nagaveni, H.C. and Muthukrishnan, R. (2011). Pests and diseases of sandalwood plants in nurseries and their management. Working Papers of the Finnish Forest Research Institute 11 pp 69-75.

- Roxburgh, W. 1820. Santalum. Pp. 442-445. In: Carey, W. (ed.) Flora indica, vol I. Mission, Calcutta, India.Sen-Sarma, P.K. 1975. Spike disease of sandal-a yellows type disease. In: Pest and diseases of fast-growing hardwoods. 2nd FAO/IUFRO World Technical conference on Forest Disease and Insects, New Delhi, India. FOR:FAO/IUFRO/DI/75/16, Rome, Italy.

- Sahai A. & Shivanna K.R. (1984). Seed germination, seedling growth and haustorial induction in *Santalum album*, a semi-root parasite. Proceedings of the Indian Academy of Sciences 93: 571 – 580

- Savill, P. Evans, J. Auclair, D. Falk, J. (1997). Plantation Silviculture in Europe. Oxford University Press. Oxford. ISBN 0-19-854909-1

- Shashidhara G, Hema MV, Koshy B, Farooqi AA. Assessment of genetic diversity and identification of core collection in sandalwood germplasm using RAPDs. J Hort Sci Biotech. 2003;78:528–536.

- Shetty, H.R. 1977. Is sandal exotic? Indian Forester 103(5): 359-367.

- Sinha, R.L. 1961. Sandalwood in Bundelkhand Forest Division, Uttar Pradesh. Indian Forester 87(10): 590-597.

- Silva, Jaime A. Teixeira da *et al.* "Sandalwood: basic biology, tissue culture, and genetic transformation." Planta 243 (2015): 847-887.

- Singh, B.K. and Shankar, P. 2007. Status of sandalwood (*Santalum album* L.) in Karnataka. Pp. 9-13. In: Gairola, S., Rathore, T.S., Joshi, G., Arun Kumar, A.N. and Aggarwal P.K. (eds.) Conservation, Improvement, Cultivation and Management of Sandal (*Santalum album* L.). IWST, Bangalore.

- Sprague, T.A. and Summerhayes, V.S. 1927. *Santalum, Eucarya,* and Mida. Bull. Miscellaneous Information Royal Botanic Gardens, Kew 5: 193-202.

- Srimathi R.A. and Sreenivasaya M., (1962) Occurrence of endopolyploidy in the haustorium of *Santalum album* Linn. Current Science,; 31: 69-70.

- Srimathi R.A. & Rao P.S. (1969). Accelerated germination of sandal seed. Indian Forester 95: 158 – 159

- Srimathi, R.A., Kulkarni, H.D. and Venkatesan, K.R. 1995. Recent Advance in Research and Management of Sandal (*Santalum album* L.) in India. Associated Publishing Co., New Delhi, 416 p.

- Srinivasan, V.V., Sivaramakrisnan, V.R., Rangaswamy, C.R., Ananthapadmanabha, H.S. and Shankarnarayana, K.H. (1992). Sandal (*Santalum album* L.). ICFRE, Dehra Dun.

- St. John, H. 1947. The history, present distribution, and abundance of sandalwood on O'ahu, Hawaiian Islands: Hawaiian plant studies 14. Pacific Science 1: 5-20.

- Struthers, R., Lamong, B.B., Fox, J.E.D., Wejesuriya, S.R. and Crossland, T. 1986. Mineral nutrition of sandalwood (*Santalum spicatum*). Journal of Experimental Biology 37(182): 1274-1284

- Subasinghe, Upul. (2013). Sandalwood Research: A Global Perspective. Journal of Tropical Forestry and Environment. 03. 1-8. 10.13140/2.1.2548.5445.

- Suma, T.B and Balasundaran, M. 2003. Isozyme variation in five provenances of *Santalum album* in India. Aust Jour of Botany 51(3): 243-249.

- Sundararaj, R. and Sharma, G. 2010. Studies on the floral composition in the six selected provenances of Sandal (*Santalum album* Linnaeus) of South India. Biological Forum-An International Journal 2(2): 73-77.

- Thirawat, S. 1955. Spike disease of sandal. Indian Forester 81: 804.

- Troup, R.S. 1921. The silviculture of Indian tree. Vol.III. Clarendon Press, Oxford.

- Tuyama, T. 1939. On *Santalum boninense*, and the distribution of the species of *Santalum*. J. Jap. Bot. 15: 697-712.

- Uma Shaanker, R., Ganeshaiah, K.N. and Nageswara Rao, M. 2000. Conservation of sandal genetic resource in India: problems and prospects. In: International conference on science and technology for managing plant genetic diversity in the 21st Century, Kuala Lumpur, Malaysia.

- van Dijk, J.A. 1995. Indigenous and Introduced Soil and Water Conservation in Sudan. Waterlines, 13(4):19-21.

- Venkatesan, K.R. 1980. A fresh look at management of Sandal. Pp. 1101-1108. In: Proceedings of Second Forestry Conference Vol. II, FRI and Colleges, Dehra Dun.

- Venkatesan, K.R., Srimathi, R.A. and Kulkarni, H.D. 1995. Survey population. Pp. 3-52. In: Srimathi, R.A., Kulkarni, H.D. and Venkatesan, K.R. (eds.) Recent Advance in Research and Management of Sandal (*Santalum album* L) in India. Associated Publishing Company, New Delhi.

- Woodall G.S. (2004). Cracking the woody endocarp of *Santalum spicatum* nuts by wetting and rapid drying improves germination. Australian Journal of Botany 52: 163 – 169.

- Yani, Septiani, Elizabeth Lukas, Dodi Andriadi S, Esther Martha Joseph and Christian Mocka (2010). Techical report inventory ofsandalwood treesat Timor Tengah Selatan District. Forestry and Estate Crop Service of Timor Tengah Selatan District. ITTO publication 49p.

- Young, A., Boshier, D. and Boyle, T. (2000). Forest Conservation Genetics: Principles and Practice. Edited by A. Young, D. Boshier and T. Boyle. CSIRO Publishing, Collingwood VIC, Australia pp. 13

- Zhang Xin-Hua, Jaime A. Teixeira da Silva, and Guo-Hua Ma (2010). Karyotype analysis of *Santalum album* L.;Vol. 63, no. 2: 142-148

- Zobel, B. J., & Talbert, J. (1984). Applied Forest Tree Improvement (505 p). New York: John Wiley.

www.ingramcontent.com/pod-product-compliance
Lightning Source LLC
Chambersburg PA
CBHW040125270326
41926CB00001B/16